Lecture Notes in Computer Science   5286

Commenced Publication in 1973
Founding and Former Series Editors:
Gerhard Goos, Juris Hartmanis, and Jan van Leeuwen

Maria Marinaro   Silvia Scarpetta
Yoko Yamaguchi (Eds.)

Dynamic Brain - from Neural Spikes to Behaviors

12th International Summer School on Neural Networks
Erice, Italy, December 5-12, 2007
Revised Lectures

 Springer

Volume Editors

Maria Marinaro
Università degli Studi di Salerno, Dipartimento di Fisica "E.R. Caianiello"
Via San Allende, 84081 Baronissi (SA), Italy
E-mail: marinaro@sa.infn.it

Silvia Scarpetta
Università degli Studi di Salerno, Dipartimento di Fisica "E.R. Caianiello"
Via San Allende, 84081 Baronissi (SA), Italy
E-mail: silvia@sa.infn.it

Yoko Yamaguchi
Laboratory for Dynamics of Emergent Intelligence, RIKEN Brain Science Institute
2-1 Hirosawa, Wako-shi, Saitama 351-0198, Japan
E-mail: yokoy@brain.riken.jp

Library of Congress Control Number: 2008937565

CR Subject Classification (1998): F.1, I.2, B.3, I.1.3

LNCS Sublibrary: SL 1 – Theoretical Computer Science and General Issues

ISSN   0302-9743
ISBN-10   3-540-88852-7 Springer Berlin Heidelberg New York
ISBN-13   978-3-540-88852-9 Springer Berlin Heidelberg New York

Springer is a part of Springer Science+Business Media

springer.com

© Springer-Verlag Berlin Heidelberg 2008
Printed in Germany

Typesetting: Camera-ready by author, data conversion by Scientific Publishing Services, Chennai, India
Printed on acid-free paper   SPIN: 12539864   06/3180   5 4 3 2 1 0

Preface

This volume contains invited and contributed papers presented at the 12th edition of the International Summer School on Neural Networks "Eduardo R. Caianiello," co-organized by the RIKEN BSI (Japan) and the Department of Physics of the University of Salerno (Italy).

The 12th edition of the school was directed by Maria Marinaro (University of Salerno), Silvia Scarpetta (University of Salerno) and Yoko Yamaguchi (RIKEN BSI Japan) and hosted in the Ettore Majoranca Center in Erice in Italy.

The contributions collected in this book are aimed at providing primarily high-level tutorial coverage of the fields related to neural dynamics, reporting recent experimental and theoretical results investigating the role of collective dynamics in hippocampal and parahippocampal regions and in the mammalian olfactory system.

This book is devoted to graduate students and researchers with different scientific background (including physics, mathematics, biology, neuroscience, etc.) who wish to learn about brain science beyond the boundary of their fields. Each lecture aimed to include basic guidance in each field. Topics of lectures include the hippocampus and entorhinal cortex dynamics and mammalian olfactory system dynamics, memory and phase coding, mechanisms for spatial navigation and for episodic memory function, oscillations in neural assemblies, cortical up and down states, and related topics where frontier efforts in recent decades have been successfully linked to a remarkable evolution of the field.

April 2008

M. Marinaro
S. Scarpetta
Y. Yamaguchi

Table of Contents

Neural Network Theories on Associative Memory

The Brain Computation Based on Synchronization of Nonlinear Oscillations: On Theta Rhythms in Rat Hippocampus and Human Scalp EEG

Yoko Yamaguchi

Lab. for Dynamics of Emergent Intelligence,
RIKEN Brain Science Institute, Wako, Saitama, Japan
yokoy@brain.riken.jp

Abstract. Synchronization of oscillations is widely observed in neural assemblies in the brain. To clarify the computational principle in these systems with essential nonlinearity, theta rhythms in rat hippocampus and in human scalp EEG were investigated. Recent discovery of grid cell and its contribution to place computation in the hippocampus were comprehensively understood by using a computational model with nonlinear oscillations. Human EEG theta (4–8 Hz) study also indicated the central role of synchronization for on demand module linking. Synchronization can crucially contribute to computation through unification among heterogeneous developing systems in real time. A basic principle toward intelligent system design was discussed for further study on the brain computation.

Keywords: Synchronization, theta rhythm, hippocampus, place cell, grid cell, human EEG.

1 Introduction

The brain consists of a huge number of neurons (200,000,000,000 in human) and their synaptic connections. For understanding the brain computation, it is crucial to solve the question how they can work together in parallel. Presence of synchronization of nonlinear oscillations in the brain was pointed out based on the profile of EEG power spectrum [1]. Nonlinear oscillations are typical self-organization phenomena universally observed in nonlinear non-equilibrium systems in physical and biological systems [2]. The nonlinear properties of oscillations can endow flexible and robust phenomena of "work together" where oscillators with different natural frequencies and different initial phases are pulled together into an attractor giving a collective oscillation with a common frequency and constant phase differences [3]. This phenomenon is called synchronization, entrainment or phase-locking. This is totally different from linear oscillation phenomena such as resonance. Synchronization of nonlinear oscillations is well known to cause robust organization, for example, in circadian rhythms, gait of animal locomotion.

M. Marinaro, S. Scarpetta, and Y. Yamaguchi (Eds.): Dynamic Brain, LNCS 5286, pp. 1–12, 2008.

On the other hand, in spite of the long history of EEG studies, the functional role of brain oscillation has not been clearly understood.

Here we elucidate the functional role of theta oscillation in memory and cognitive tasks based on recent development in rodent hippocampus and human EEG in cognitive task conditions. The key question below is how local processors can be coordinated in real-time computation in ever-changing environments.

2   Space Computation in Hipocampal-Entorhinal System

2.1   Theta Rhythms and Place Cells

Rodent hippocampus has been extensively studied based on the cognitive map theory by O'Keefe and Nadel [4], where firing rates of hippocampal cells encode the animal's location in an environment. These neurons are called place cells. Another important observation in rodent hippocampus [5] is theta oscillation (4–12 Hz) in local field potential (LFP), which appear during voluntary movements. The relation between rate coding for the cognitive map and LFP theta was an open question for a long time. O'Keefe and Recce [6] reported "theta phase precession" in rat hippocampus, where firing of hippocampal neuron within its place field depends on LFP theta phase. The relative phase of firing gradually advances as the rat traverses. Skaggs et al. [7] reported that phase precession is characterized also as a temporal sequence of collective place cell firing. The firing sequence of a number of place cells with overlapping place fields are compressed into each theta cycle representing the running sequence as shown in Fig.1.

We proposed a model of theta phase precession by hypothesizing the source of theta phase precession in the entrochinal cortex [8] [9]. In this hypothesis, theta phase precession is generated even in the absence of the hippocapmal associative memory; this ability is inevitable for memory formation of novel events. In the entorhinal cortex, we assumed intrinsic oscillation near theta frequency.

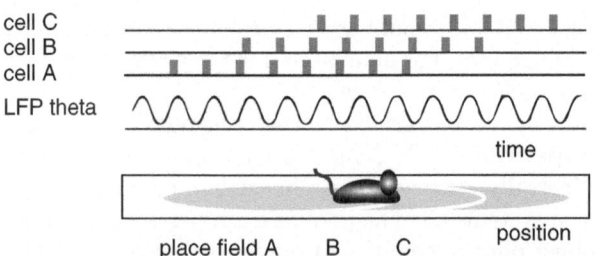

Fig. 1. Illustration of theta phase precession in rat place cells. When the rat traverses in a place field, spike timing of the place cell gradually advances relative to local field potential (LFP) theta rhythm. In a running sequence through place filed A-B-C, the spike sequence in order of A-B-C emerges in each theta cycle. The spike sequence repeatedly encoded in theta phase is considered to lead robust on-line memory formation of the running experience through asymmetric synaptic plasticity in the hippocampus.

The coupled system of EC cell and LFP theta can be described by simple equations

$$\begin{cases} \dot{\theta}_i = \omega_i + A\sin(\theta_0 - \theta_i), \\ \dot{\theta}_0 = \omega_0, \end{cases}$$

where θ_i and θ_0 are phases of EC cell and LFP, the relative phase $\theta_i - \theta_0$ is found to have an equilibrium point:

$$\theta_i - \theta_0 = \sin^{-1}\left(\frac{\omega_i - \omega_0}{A}\right).$$

respectively. ω_i and ω_0 are oscillation frequencies and A is coupling magnitude. If the intrinsic frequency ω_i is constant, $\theta_i - \theta_0$ the relative phase of unit firing to LFP theta is found constant. The relative phase is robust against any perturbation, while it does not advance. As a necessary condition for generation of gradual phase advancement in theta phase precession, the natural frequency is assumed to have slow increase during input period. That is, according to the increase of natural frequency, the relative phase gradually advances. Each firing is given by the relation of the instantaneous value of intrinsic frequency as a quasi steady state of phase locking.

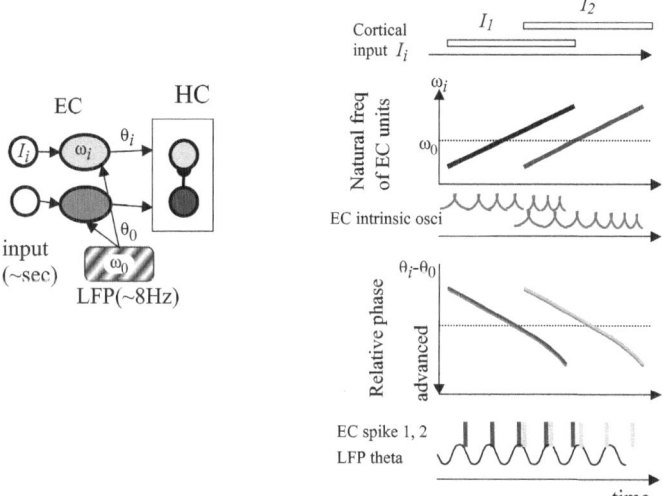

Fig. 2. A computational model of theta phase precession [9]. Theta phase precession is generated in EC cells with intrinsic oscillations. In the presence of cortical input I_i, the i-th EC unit fires at a relative phase $\theta_i - \theta_0$ that is determined by phase locking between the intrinsic oscillation (ω_i) and LFP theta ω_0. Theta phase precession emerges as gradual advancement of the relative firing phase in accordance with a slow increase of ω_i.

2.2   Discovery of Theta Phase Precession in EC Grid Cells

Recently it was found that the entorhinal neurons, giving major inputs to the hippocampus, fire at positions distributing in a form of a triangular-grid-like patterns in the environment [10]. They are called "grid cells" and their spatial firing preference is termed "grid fields". Interestingly, temporal coding of space information, "theta phase precession" initially found in hippocampal place cells were also observed in grid cells in the superficial layer of the entorhinal cortex [11], as shown in Fig. 3. A sequence of neural firing is locked to theta rhythm of LFP during spatial exploration. The presence of theta phase precession in the entorhinal cortex is in accordance with our hypothesis of theta phase precession [8] [9]. On the other hand, the grid cell discovery requires fundamental reconsideration of the hippocampal system either in its function or on its mechanism. Does this study entirely change the previous understanding of hipopcampal networks with place cells or not?

The properties of hippocampal cells and entorhinal cells are summarized in Fig.4. The new discovery in EC raises a number of interesting questions. Can the cognitive map be represented and stored in the hippocampus though these highly organized neural entities? More strictly, following questions are to be solved.

i) How is the grid field formed in the entorhinal cortex?
ii) How is phase precession generated in grid cells?
iii) Can projection of grid cells to hippocampal neurons generate place cells?
iv) Does the above bring any new principle of space computation in the brain?
v) What are computational roles of various entorhinal cells based on biophysical studies of entorhinal cells and their computational role in grid cell system?

Among on-going studies on these questions, I would like to describe our recent results on i)-iv) below.

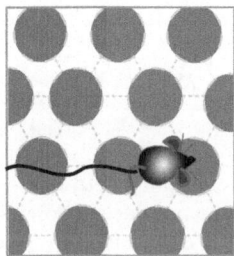 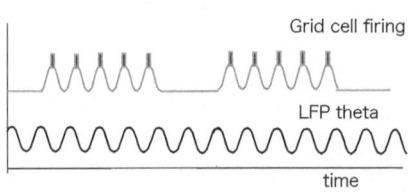

Fig. 3. (Left) A grid field of an entorhinal grid cell. (Right) Theta phase precession observed in the grid filed along a running sequence.

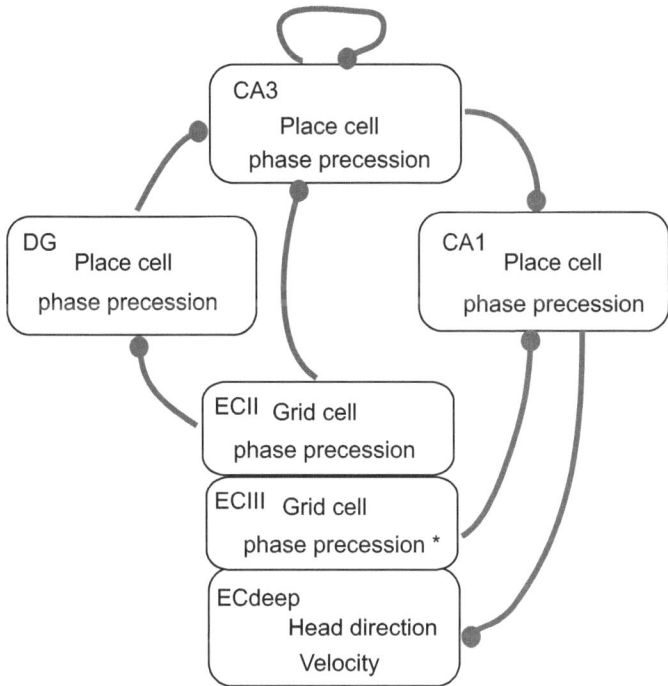

Fig. 4. A summary of neural firing properties in the entorhinal-hippocampal system, DG, CA3, CA1, the entorhinal cortex (EC deeper layer, ECII, ECIII). Theta phase precession was first found in the hippocampus, and finally in the EC superficial layer (II and III). ∗) In EC layer III, about 10 % grid cells exhibit theta phase precession.

2.3 A Model of Grid Cell Formation Based on Self-motion Cues

To obtain comprehensive understanding of space computation, we extended our former model to include grid cells in accordance with known property of entorhinal neurons including "head direction cells" which fires when the animal's head has some specific direction in the environment [12]. We demonstrate theta theta phase precession in our former model [9] naturally emerge as a consequence of the grid cell formation mechanism.

Firing rate of the ith grid cell at a location (x, y) in a given environment increases in the condition given by the relation:

$$x = \alpha_i + nA_i \cos \phi_i + mA_i \cos(\phi_i + \pi/3),$$
$$y = \beta_i + nA_i \sin \phi_i + mA_i \sin(\phi_i + \pi/3),$$

with

$$n, m = \text{integer} + r, \tag{1}$$

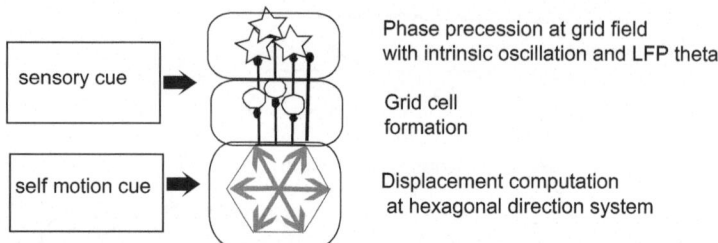

Fig. 5. Illustration of a column model for grid field computation. The bottom layer consists of local path integration module with a hexagonal direction system. The middle layer associates output of local path integration and sensory cue in a given environment. The top layer consists of a set of grid cells whose grid fields have a common orientation, a common spatial scale and complementary spatial phases. Phase precession is generated at the grid cell at each grid field.

where ϕ_i, A_i and (α_i, β_i) denote one of angles characterizing the grid orientation, a distance of nearby vertices, and a spatial phase of the grid field in an environment. The parameter r is less than 1 representing the size of a field with high firing rate.

The computational goal to create a grid field is to find the region with $n, m =$ integer $+ r$.

We hypothesize that the deeper layer of the entorhinal cortex works as local path integration systems by using head direction and running velocity. The local path integration results in a variable with slow gradual change forming a grid field. This change can cause the gradual phase shift of theta phase precession in accordance with the phenomenological model of theta phase precession by Yamaguchi et al. [13].

As shown in Fig. 4, the entorhinal layer includes head direction cells in the deeper layer and grid cells in the superficial layer. Cells with theta phase precession can be considered as stellate cells. The set of modules along vertical direction form a kind of functional column with a direction preference as shown in Fig.5. These columns form a hypercolumnar structure with a set of directions.

The local path integration module consists of six units. During animal's locomotion with a given head direction and velocity, each unit integrates running distance in each direction with an angle dependent coefficient.

Computation of animal displacement in given directions in this module is illustrated in Fig, 6. The maximum integration length of the distance in each direction is assumed to be common in a module, corresponding to the distance between nearby vertices of the subsequently formed grid field. This computation gives (n, m).

Computational results of local path integration are projected to next module in the superficial layer of the entorhinal cortex, which has multiple sensory inputs in a given environment. The association of path integration and visual

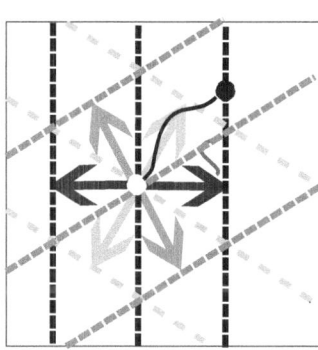 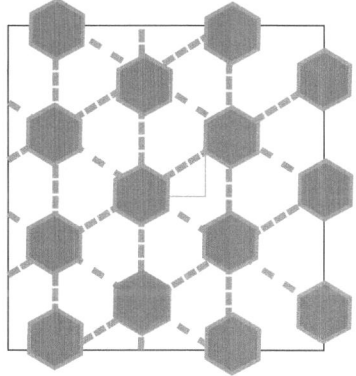

Fig. 6. (Left) Illustration of local path integration in a hexagonal direction system. Animal locomotion in from the white circle to the black circle is computed by individual vector unit among six vectors to give a position measure. (Right) A grid field computed by the module.

cues results in the relative location of path integration measure (α_i, β_i) in the module.

The input of the parameter (n, m) and (α_i, β_i), to a cell at next module, at the top part of the module, can cause theta phase precession in the assumption given by the former model [13] [8]. The natural increase in frequency is expected to emerge by the input of path integration at each vertex of a grid field.

Simple mathematical formulation of the above model is given below. The locomotion of animal is represented by a displacement distance in a given time interval ΔR and its direction ϕ_d. The latter coincide with the head direction in a small time interval condition. An elementary vector at a column of in local path integration system has a vector angle ϕ_i and its modulus length A_0. The output of the j-th hexagonal vector system D_j is given by

$$D_j = \prod_{j=1}^{6} I_j(\phi_i) \tag{2}$$

with

$$I_j(\phi_i) = \begin{cases} 1 & \text{if } -r < S_j(\phi_i) < r, \\ 0 & \text{otherwise} \end{cases}$$

and

$$S_j(\phi_i) = \Delta R \cos(\phi_i - \phi_d) \quad (S \bmod A).$$

where r represent the field radius.

Through association with visual cues, spatial phase of the grid is determined. (Details are not shown here.)

The term Eq. (2) from the middle layer to the top layer gives on-off regulation and also a parameter with gradual increase in a grid field.

Dynamics of the membrane potential V_{jk} of the cell at the top layer can be described by a biophysical equation as follows.

$$\frac{d}{dt}V_{jk} = f(V_{jk}, t) + I(\{S_j\}) + I_{\text{theta}}, \tag{3}$$

where f is a function of time-dependent ionic currents. The second and last terms respectively representinput from hexagonal direction vector system and a sinusoidal current representing theta oscillation of inhibitory neurons. In a proper dynamics of f, the second term in the right had side gives activation of the grid cell oscillation and gradual increase in its natural frequency. According to our former results by using a phenomenological model [8], the last term of theta currents leads phase locking of grid cells with gradual phase shift. This realizes a cell with grid field and theta phase precession. We also tested Eq. (3) by using biophysical equation for several types of neurons. We obtained similar phase precession with stellate cell model. One important property of stellate cell is the presence of sub threshold oscillations, while synchronization of this oscillation can be reduced to a simple behavior of the phase model. Thus, the mechanism of phenomenological model [8] is found to endow comprehensive description of phase locking of complex biophysical neuron models.

2.4   From Grid Cells to Place Cells

Our computational model of formation of grid field was proposed based on local path integration. This assumption was found to give theta phase precession within the grid field. This computational mechanism does not always need an assumption of learning in repeated trials in an environment but enables instantaneous spatial representation.

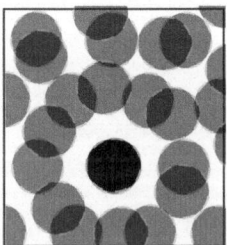

Fig. 7. An example of superposition f firing rate of two grid cells with the same space scale and the different direction. One field with good matching between two grid fileds is surrounded by broad distribution of weak firing rate. When phase precession in grid cells and coincidence detection in a hippocampal cellare considered, the single filed with good mathig is selectively transmitted to the hippocampal neuron, while others are excludedd.

The last question in this section is computation of place cells based on grid cells. Projection of EC II cells to a cell either in the DG or in the CA3 is given by a set of grid cells. That is, hippocamal neuron received superposition of grid fields as shown in Fig.7. In superposition of firing rates, some part happen to be found as a single place filed, while it is in general associated with delocalized firing, disrupting place filed segregation. We have analyzed this problem by considering theta phase precession at each grid cell [19]. If a hippocampal neuron is a linear integrator of entorhinal spikes, the place field does not appear. On the other hand, if a hippocamal neuron is assumed to be coincidence detector, a set of in-phase spikes are selectively integrated to give a localized activity in a place filed. The coincidence detection of phase along theta phase precession namely leads inheritance of theta phase precession from entorhinal cells to hippocampal cells inseparably with formation of place field.

Our consideration above is summarized in Fig. 8. We find fascinating nature of space representation in cognitive map system in a highly organized space-time domain.

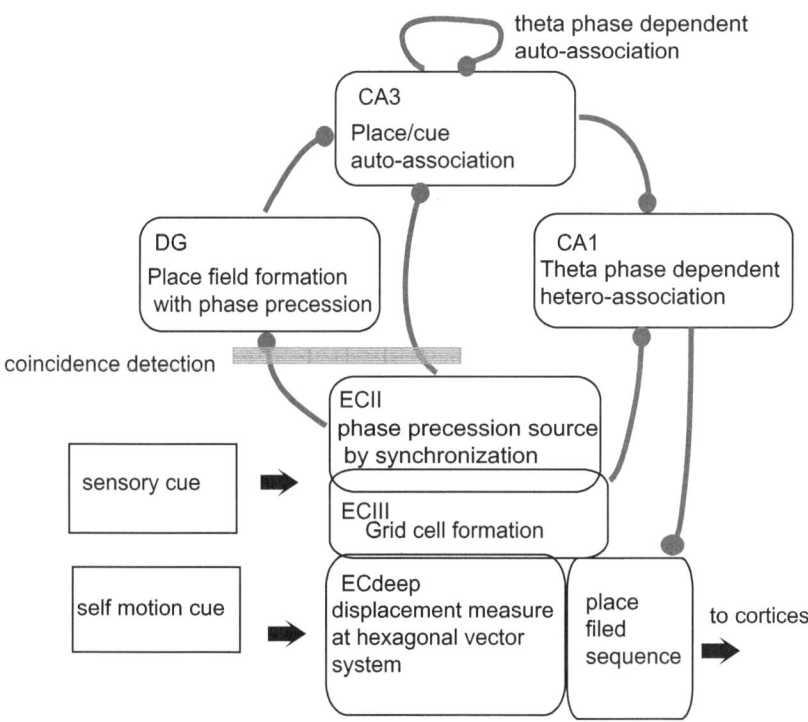

Fig. 8. Summary on neural computation of space and its memory formation in entorhinal-hippocampal system

3   EEG Synchronization as a Key for Elucidation of Human Intelligence

A physiological study in olfactory system by Walter Freeman was one of frontier of neural oscillation for cognitive functions in the brain. Current studies on brain dynamics in various levels more and more reveal the dynamical brain coordinated by synchronization of oscillation. On the other hand, we are still far from the answer on computational mechanism for cognitive functions.

Our synchronization network model of visual pattern recognition was first proposed in 1985 (Shimizu et al.) [14]. The model was devoted to enlighten the necessity of information flow between visual image and concept memory to lead dynamical binding among relevant features, which was followed by report of Gamma synchronization in cat visual cortex by Singer and Gray in 1989 [15]. The extended model [16] [17] exhibits pattern recognition through figure-ground separation, while recognition of complex scenes often fails to arrive a fixed point It suggests any additional control system in the brain.

Ishihara et al reported human EEG theta in 1972 with a task of IQ test [18]. Theta rhythm was typically increased in task-dependent manner at frontal midline regions. It is called fm theta for short. The current source of fm thata was estimated at the anterior cingulated cortex or distributed regions in frontal medial wall. Importantly, this region is known to concern with central executive functions including attention, working memory, future monitor, performance monitor etc. Because of the task property of central executive functions, it should be important to have ability of dynamical linking according to task demands in indefinite situations.

We hypothesized that fm theta is related with dynamical linking among various associative areas and monitor areas in frontal medial regions. To test this, we developed EEG index associated entire brain activities by using simultaneous

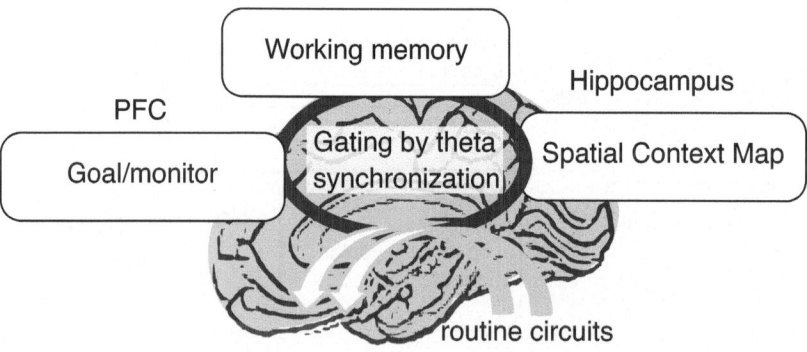

Fig. 9. Brain networks for the control of action-perception based on theta synchronization. Dynamical linking of working memory system as well as hippocampal memory system enables regulation of sensor-motor system (or thinking as an inner motor behavior) in areal-time adoptive manner.

EEG and fMRI measurements. In the task of mental calculation, we successfully found that monitor areas in the frontal medial wall are dynamically and alternately linked with other cortical modules such as working memory regions associated with distant theta synchronization increase. This study suggests that brain computation for intelligence in general uses the principle of synchronization of the slow oscillation. A generalized view of the brain computation is illustrated un Fig. 9.

4 Concluding Remarks

Here we elucidated a computational model of space computation in rodent entorhinal-hippocampal system. It gives very strong evidence on the crucial role of synchronization in the brain for parallel computation. Sato elucidates the question in human hippocampus memory in this book.

Toward the computational theory of intelligence, human EEG evidence was shortly reviewed. To our happiness, recent development in animal and human is huge enough to construct a concrete view for the intelligent machine. It could be available for robotics or intelligent system design in general.

The brain is not only synchronization - - -but the brain cannot be solved without synchronization.

Acknowledgments

I would like to express acknowledgement to all colleagues in Laboratory for Dynamics of Emergent Intelligence, RIKEN BSI for their hard efforts in collaboration, and to Drs. Marinaro, Scarpetta, all lecturers and participants for giving me fascinating opportunity of the school in Erice. I also would like to express my appreciation to Dr. Sato for his kind help in manuscript arrangement, and finally to Ms. Chiba for her powerful management on the organization of the school.

References

1. Wiener, N.: Non-linear problems in random theory. MIT Press, Cambridge (1966)
2. Haken, H.: Synergetics; An Introduction non-equilibrium phase transitions and self-organization in physics, chemistry and biology. Springer, Heidelberg (1977)
3. Winfree, A.: The Geometry of Biological Time, 2nd edn. Springer, New York (2001)
4. O'Keefe, J., Nadel, L.: The hippocampus as a cognitive map. Clarendon Press, Oxford (1978)
5. Vanderwolf, C.H.: Hippocampal electrical activity and voluntary movement in the rat. Electroen Clin. Neuro. 26, 407–418 (1969)
6. O'Keefe, J., Recce, M.L.: Phase relationship between hippocampal place units and the EEG theta rhythm. Hippocampus 3, 317–330 (1993)
7. Skaggs, W.E., McNaughton, B.L., Wilson, M.A., Barnes, C.A.: Theta phase precession in hippocampal neuronal populations and the compression of temporal sequences. Hippocampus 6, 149–172 (1996)

8. Yamaguchi, Y., McNaughton, B.L.: Nonlinear dynamics generating theta phase precession in hippocampal closed circuit and generation of episodic memory. In: Usui, S., Omori, T. (eds.) The Fifth International Conference on Neural Information Processing (ICONIP 1998) and The 1998 Annual Conference of the Japanese Neural Network Society (JNNS 1998), vol. 2, pp. 781–784. IOS Press, Kitakyushu (1998)
9. Yamaguchi, Y.: A theory of hippocampal memory based on theta phase precession. Biol. Cybern. 89, 1–9 (2003)
10. Fyhn, M., Molden, S., Witter, M., Moser, E.I., Moser, M.B.: Spatial representation in the entorhinal cortex. Sience 305, 1258–1264 (2004)
11. Hafting, T., Fyhn, M., Moser, M.B., Moser, E.I.: Phase precession and phase locking in entorhinal grid cells. Program No. 68.8. 2006 Neuroscience Meeting Planner. Society for Neuroscience, Atlanta (2006)
12. Muller, R.: A quarter of a century of place cells. Neuron 17(5), 813–822 (1996)
13. Yamaguchi, Y., Sato, N., Wagatsuma, H., Wu, Z., Molter, C., Aota, Y.: A unified view of theta-phase coding in the entorhinal-hippocampal system. Current Opinion in Neurobiology 17, 197–204 (2007)
14. Shimizu, H., Yamaguchi, Y., Tsuda, I., Yano, M.: Pattern recognition based on horonic information of vision. In: Haken, H. (ed.) Complex systems – operational approaches in neurobiology, phisics, and computers, pp. 225–239. Springer, Heidelberg (1985)
15. Gray, C.M., Koenig, P., Engel, K.P., Singer, W.: Oscillatory responses in cat visual cortex exhibit inter-columnar synchronization which reflects global stimuli properties. Nature 338, 334–337 (1989)
16. Yamaguchi, Y., Shimizu, H.: Pattern recognition with figure–ground separation by generation of coherent oscillations. Newral Network 7, 49–63 (1994)
17. Hirakura, Y., Yamaguchi, Y., Shimizu, H., Nagai, S.: Dynamic linking among neural oscillators leads to flexible pattern recognition with figure–ground separation. Neural Network 9, 189–209 (1996)
18. Ishihara, T., Yoshii, N.: Multivariate analytic study of EEG and mental activity in juvenile delinquents. Electroencephalogr. Clin. Neurophysiol. 33, 71–80 (1972)
19. Molter, C., Yamaguchi, Y.: Entorhinal theta phase precession sculpts dentate gyrus place fields. Hippocampus (2008), doi:10.1002/hipo.20450

Theta Phase Coding in Human Hippocampus: A Combined Approach of a Computational Model and Human Brain Activity Analyses

Naoyuki Sato

Lab. for Dynamics of Emergent Intelligence
RIKEN Brain Science Institute, Saitama 351-0198, Japan
satonao@brain.riken.jp

Abstract. The hippocampus is known to maintain episodic memory in humans, but its neural mechanism is an open question. The rat hippocampus is well investigated and known to have a particular neural synchronization between firing and local field potential (LFP) called 'theta phase precession' that encodes spatio-temporal input sequences to the hippocampus. According to anatomical similarity between the rat and human hippocampus, similar dynamics are expected in the human hippocampus, while it is still unknown if the phase precession dynamics could contribute to memory formation of complex information in episodic memory. In this paper, we evaluate the human hippocampal dynamics by using a combined approach of a computational model and human brain activity analyses. A series of computational and experimental analyses provides integrative understanding of human hippocampal dynamics and its function.

1 Introduction

The neural synchronization dynamics play a key role for bridging neuron-level and behavior-level dynamics (Fig. 1). One good example is locomotion pattern generation in the lamprey [1]. In the lamprey, motoneurons in the spinal cord are synchronously activated and the synchronization is associated with locomotion pattern. The synchronization dynamics are clearly necessary for avoiding muscle damage and for generating economical locomotion patterns. Interestingly, neural synchronization dynamics are also associated with information processing of environmental changes (e.g. water flow) that would be achieved by the spinal cord. Another example is feature binding in the visual cortex [2], where two neurons with close receptive fields are shown to be synchronously activated in relation to coherence of visual stimuli. This indicates that the neural synchronization would be associated with perceptual grouping and could solve the 'binding problem' of multi-modal visual features, such as color, motion, shape, etc. Again neural synchronization dynamics are a bridge between neuron-level and behavior-level dynamics.

In this paper, we will focus on the dynamics of neural synchronization in the hippocampus. We begin by introducing a basic mechanism of neural synchronization

M. Marinaro, S. Scarpetta, and Y. Yamaguchi (Eds.): Dynamic Brain, LNCS 5286, pp. 13–27, 2008.

and propose a computational model of the human hippocampus. We will show how the computational model integrates experimental data consisting of human memory recall, eye movement, electroencephalography (EEG) and hippocampal BOLD (Blood Oxygenation Level-Dependent) signal. The importance of the combined approach of a computational model and human experiment will be discussed.

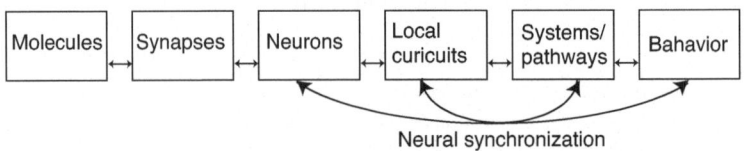

Fig. 1. Levels of organization in the brain. Neural synchronization dynamics play a key role for integrative understanding of multi-level brain dynamics.

2   Dynamics of Neural Oscillation in the Hippocampus

2.1   The Hippocampus

The hippocampus is part of the limbic system and is associated with memory function. Rodents and primates have hippocampi with a similar anatomical structure [3] that is characterized by a closed circuit; cortical inputs enter the superficial layer of the entorhinal cortex and sequentially project to the dentate gyrus, CA3, CA1, the subiculum and the deeper layer of the entorhinal cortex. The CA3 includes massive recurrent connections that are expected to implement an associative memory [4]. The cortico-hippocampal connectivity is well studied and shows that the hippocampus is reciprocally connected to wide neocortical regions through the parahippocampal region [5].

In humans, a patient with hippocampal damage, H.M. [6], clearly demonstrates the importance of the hippocampus to episodic memory [7] (for example, the menu of today's breakfast etc.) Hippocampal damage leads to difficulty in forming new memories, while old memories (two years before the damage) , procedural memory, language skills and IQ scores remain normal after the damage.

2.2   Theta Oscillation in the Hippocampus

Various behaviors are known to reflect specific frequency oscillations of local field potential (LFP) and EEG. In humans, delta oscillations (1–4 Hz) appear during sleep, theta oscillations (4–8 Hz) are associated with memory function, alpha oscillations (8–12 Hz) strongly appear during eye closing, beta-band oscillations (12–20 Hz) are associated with attentional processing, and gamma-band oscillations (20–100 Hz) are known to relate to feature binding. In the rat hippocampus, LFP theta power clearly increases during walking [8], and a similar property is also reported in humans [9]. The hippocampal theta dynamics are considered to be associated with information processing of spatial navigation.

The hippocamapal theta oscillation is also known to relate to odor encoding: The hippocampal LFP theta is synchronized with the physical movement of odor sniffing. Importantly the LFP coherence at the sniffing frequency between the olfactory bulb and the hippocampus is known to increase during successful odor encoding [10]. This evidence suggests that hippocampal theta oscillation plays a key role in memory function.

2.3 Modeling Neural Synchronization Dynamics

How can we understand the dynamics of neural oscillations? One of the simplest models of neural oscillation is a linear oscillator described by the second order differential equation (Fig. 3a). In the linear oscillator, an exact sine wave appears and the amplitude of the oscillation depends on the initial state of the oscillator. When a dumping factor is included (Fig. 3b), the amplitude of the oscillation decreases. In terms of the robustness of the oscillation, these are not a likely model of neural oscillation. When a negative damping factor is assumed around $x = 0$ (Fig. 3c), a stable limit cycle appears independent of the initial state. This description is known as a van der Pol oscillator that is a typical nonlinear oscillator. The van der pol oscillator is also adapted for neural dynamics that show a resting state, excitability and a periodic oscillation state. For example, the Fitzhugh-Nagumo equation was proposed to describe oscillations of a spike train [11], and the phase equation [12] was also proposed to describe the same dynamics in the phase plane by using a small set of parameters.

The nonlinear oscillators could perform real-time synchronization and phase locking. When two nonlinear oscillators with different oscillatory phases are coupled, the oscillations could be immediately synchronized (Fig. 3d). On the other hand, linear oscillators are difficult to synchronize (Fig. 3e). Even when these nonlinear oscillators have slightly different native frequencies, the oscillators are also synchronized with a constant phase difference. The dynamics of nonlinear oscillators would be essential for the understanding of neural oscillations.

2.4 Memory Encoding by Theta Phase Precession

O'Keefe et al. [13] discovered an interesting relationship between place cell's firing and LFP theta in the rat hippocampus. In this phenomenon, firing of the place cells are synchronized with LFP theta, and these firing phases with LFP theta are gradually advanced as the rat passes through the place field. This phenomenon is called 'theta phase precession.' Results of multiunit recording further showed that different cells have an independent phase, thus the place fields' sequence is represented in a firing sequence is repeated in every theta cycle, as a temporally compressed representation [14]. The time scale of the phase pattern is similar to the asymmetric time-window of the Hebb rule, thus the pattern could contribute to the formation of the temporal sequence memory.

What neural dynamics generate the theta phase precession? There are many proposals based on the interference of two oscillations [13] [15], activity propagation by asymmetric connections [16] [17] [18], asymmetric rate code and

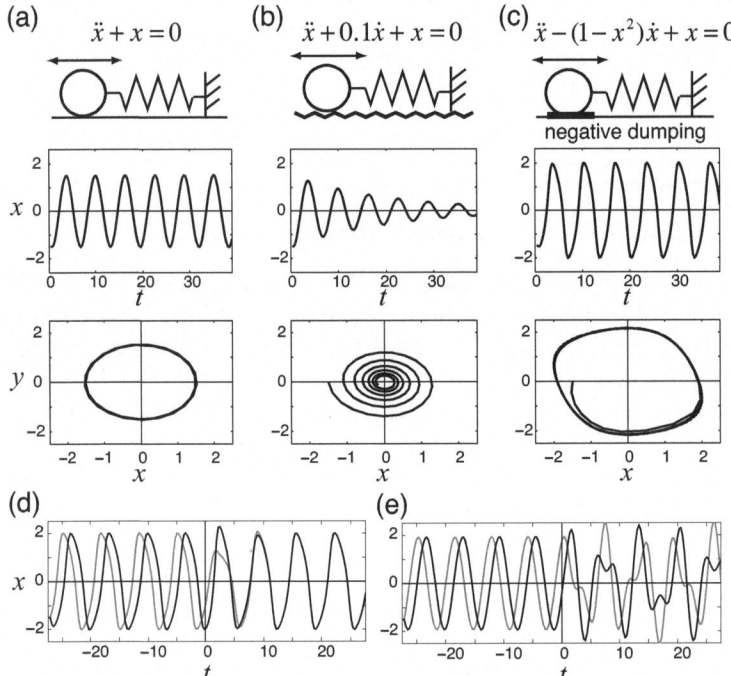

Fig. 2. Descriptions of neural oscillators. (a) A linear oscillator. Amplitude of the oscillation depends on the initial state. (b) An oscillator with a damping factor. (c) Van der Pol oscillator. Stable oscillation appears independent to the initial condition. (d) Coupling of two nonlinear oscillators. The oscillators are connected at $t = 0$. The oscillators are immediately synchronized. (e) Coupling of two linear oscillators. The oscillators are not synchronized.

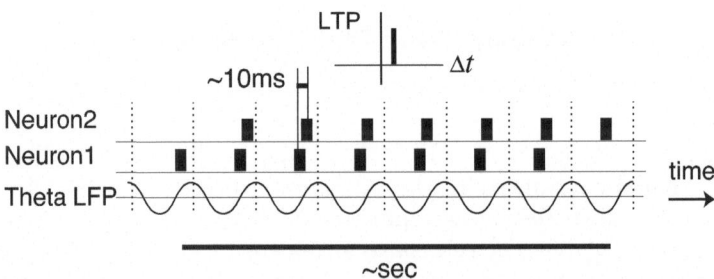

Fig. 3. Theta phase precession. The firing phases of a place cell advance as the rat passes through its place field. Interestingly, sequential activation of place cells is temporally compressed in sequential firing of cells in every theta oscillations. The time scale of the firing pattern is similar to the Hebb rule with an asymmetric time window, thus phase precession is expected to contribute to the synaptic plasticity.

oscillatory inhibition [19] and somatic inhibition-dendric excitation interference [20], while Yamaguchi [21] proposed that the theta phase precession is generated by neural synchronization dynamics in the entorhinal cortex. In this model, a gradual increase of native frequency of the entorhinal cell is assumed as an accumulation of cortical input. The synchronization between cell's activation and LFP theta results in a robust generation of phase precession. It is also important that the generation of phase precession in the entorhinal cortex enables memory storage of novel cortical inputs in the CA3 network with a Hebb rule with an asymmetric time window [21]. Results of computer simulation showed that entorhinal phase precession is an efficient dynamic for storing temporal sequence in various time scales experienced only one time [22] and for forming spatio-temporal memory [23] and the cognitive map [24]. Recently entorhinal phase precession is experimentally found in the grid cells [25] and its computational role in spatial navigation is further discussed [15].

3   A Model of Theta Phase Precession in Human Hippocampus

3.1   Object-Place Memory

To evaluate the dynamics of the human hippocampus, 'object-place memory,' consisting of what and where memory contents, is often used as an experimental model of the episodic memory. In the object-place memory task, the subject is asked to remember certain objects and their locations in the environment, and after a delay period, the subject is asked to recall the objects and these locations. When the hippocampus is damaged, the subject has great difficulty performing the task [26] [27]. This clear dependence of the object-place memory on the hippocampus allows for evaluation of hippocampal memory dynamics. The object-place memory procedure is also applied to monkeys [28] and rats [29] and also known to be associated with the hippocampus.

Another advantage of the object-place memory paradigm is an anatomical relevance: the hippocampus receives a convergent projection of what and where information from dorsal and ventral visual pathways through the parahippocampal region, respectively [30]. A human fMRI (functional Magnetic Resonance Imaging) study also demonstrated object and scene selectivities in these pathways [31]. This anatomical evidence is in agreement with the functional demands of the hippocampus forming object-place memories. Taken together, the object-place memory paradigm allows for evaluation of memory processing in the human hippocampus.

3.2   A Computational Model of Object-Place Memory

In the rat hippocampus, place cells are known to be selectively activated by the rat's location in the environment, while view cells in the primate hippocampus are known to be selectively activated by eye fixation location in the environment [32]. These differences are considered to be derived from the difference of the visual input angle between primates and rodents [33]. Together with the anatomical evidence of dorsal and ventral visual inputs, the object and scene information would be modeled by inputs from the central and peripheral visual fields that lead to small and large overlaps within those inputs [34]. This assumption is also in line with a model proposed by Rolls [35] where object-place memory is modeled with a continuous and discrete attractor.

In human object-place memory tasks, object-place memory is known to be formed during a short encoding period, and this property is also important to the model of the episodic memory. What neural mechanism can realize the instantaneous formation of object-place memories? One neural mechanism is the theta phase precession dynamic that can realize the on-line memory encoding of a temporal sequence [21] [22]. Based on theta phase coding, the authors proposed a computational model of object-place memory in the hippocampus [34]. In this model, the following scenario is hypothesized: First, multiple object-place associations are encoded by a visual input sequence consisting of object information with small overlaps and scene information with large overlaps. Second, the sequence is translated to theta phase coding at the entorhinal cortex, and finally stored into CA3 connections that form a cognitive map for object-scene associations.

By using computer experiments, the above hypothesis was tested in one-dimensional [34] and two-dimensional visual environments [36]. Fig. 4 shows the result of two-dimensional object-place associations. The visual environment included four objects (Fig. 4a) that are encoded by a randomly saccadic input sequence. According to smaller overlaps within object information, object inputs have shorter durations than scene inputs on average (Fig. 4b). In theta phase coding (Fig. 4c), these short duration inputs (i.e., shorter than the time scale of phase precession) are always translated to early phase firings, therefore a robust phase difference appears between object and scene units (Fig. 4d). After a few seconds encoding period, object-scene and scene-scene asymmetric connections are formed according to the Hebb rule with an asymmetric time window (Fig. 4e). Interestingly, asymmetric connections also appear between larger and smaller spatial scales of scene information. In summary, object-place memories are formed as a hierarchical structure of the network.

One functional advantage of the hierarchical network is selective activation of the object-place memories. When the top of the network is activated, a set of object-place associations appears simultaneously, and individual sets of object-place association appears sequentially one by one (Fig. 5a). On the other hand, an initial activation at the middle of the hierarchical structure results in the recall of a few object-place associations (Fig. 5b). Such a selective recall would be necessitated by the hippocampus that maintains huge memory contents appearing in the episodic memory. In a traditional associative network, the memory is represented by a fixed-point attractor, therefore it is difficult to realize such a selective recall.

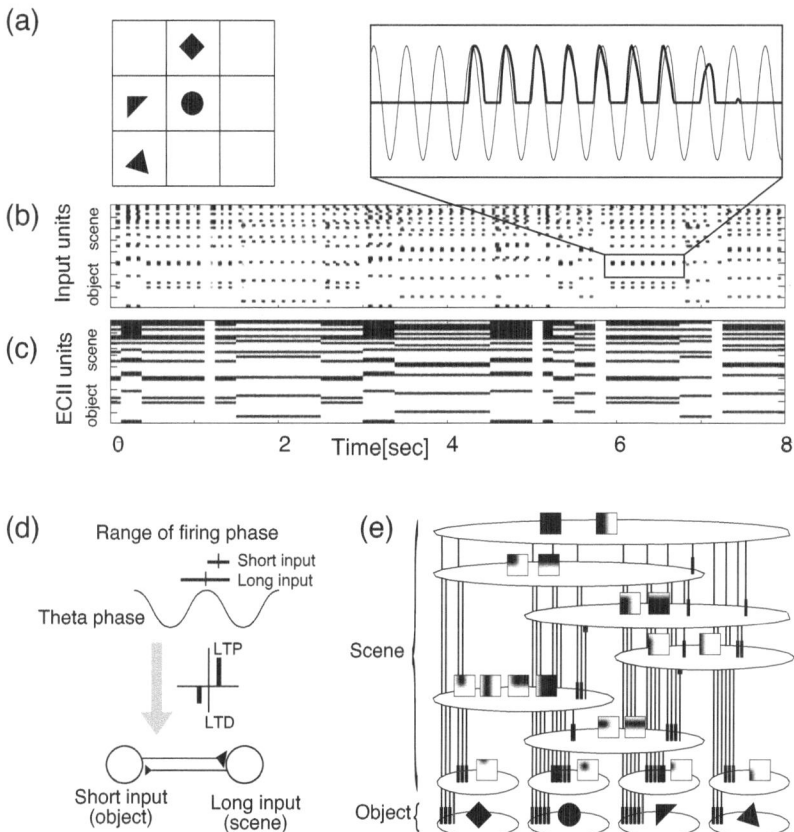

Fig. 4. Object-place memory encoding by theta phase precession. (a) Visual environment. (b) Input sequence consisting of small-overlapped object information and large-overlapped scene information. (c) Theta phase precession. (d) A schematic representation of average phase relationship between object and scene inputs. (e) Graphical representation of a resultant CA3 connection. Object-scene and scene-scene asymmetric connections are formed by phase difference of these inputs and a Hebb rule with an asymmetric time window.

The current object-place hierarchical network would have a link to the human cognitive map that is psychologically known to possess a hierarchical structure among landmarks: Smaller scale spatial relationships of landmarks (e.g., Denver v.s. Cincinnati) are influenced by larger scale spatial relationships of landmarks (e.g., Colorado vs. Ohaio) [37]. Also in laboratory environments, the similar memory structure is formed in an object-place memory paradigm [38]. Neural representation of those large-scale spatial memories is an open question, while the current results suggest that spatial inclusion ('part-of') relationships are encoded by asymmetric connections realizing a selective recall in the hippocampus.

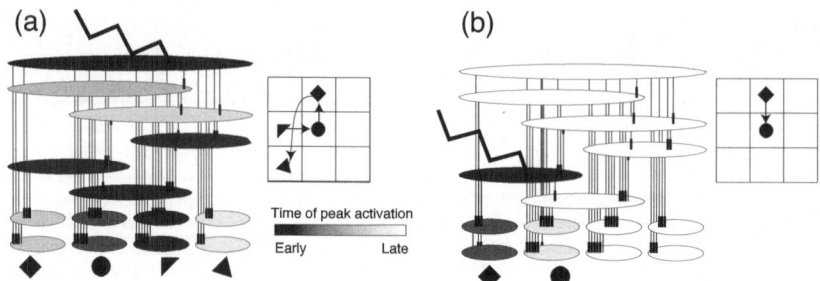

Fig. 5. Results of selective recall in a hierarchical network. (a) A recall of all four object-place associations by an initial input to the top of the hierarchical network. (b) A recall of two object-place associations evoked by an initial input of the middle of the hierarchical network.

4   Combined Approach of Computational Model and Human Brain Activity Analyses

The hierarchal network of object-place memory has an advantage in memory recall, while it remains unknown if the neural encoding of the theta phase coding and the resultant hierarchical network of the object-place memory really exist in the human brain. In the following section, we will show a series of experimental evaluations (Fig. 6) of theoretical predictions by using human non-invasive experimental measurements, scalp EEG, eye movement, human recall performance and simultaneous EEG-fMRI.

4.1   Scalp EEG

Scalp EEG that dominantly detects neocortical LFP, is difficult to associate with hippocampal EEG theta activity, while it is still important to evaluate scalp EEG during object-place memory encoding that would be implemented by cooperation among the visual and oculo-motor systems and the hippocampus. According to the theta phase coding theory, hippocampal EEG theta power is expected to increase during object-place encoding in relation to subsequently successful recall. We evaluated whether the scalp EEG theta power during object-place encoding is correlated with subsequent memory recall [39].

In the experiment, we measured 58-ch scalp EEG signals, 4-ch electrooculography (EOG) signals and eye movement from 12 participants while performing an object-place memory task. The task consists of an 8-sec encoding of four object-place associations with familiar objects (Fig. 7a), a 10-sec random saccades task, and a 30-sec recall task on the display by using a mouse. In the analysis, EEG power during encoding of each trial that was later recalled either completely ('successful') or incompletely recalled ('failed'). According to retino-corneal electric potential of the eye, eye movements produce serious artifacts in EEG signals. These ocular artifacts were corrected

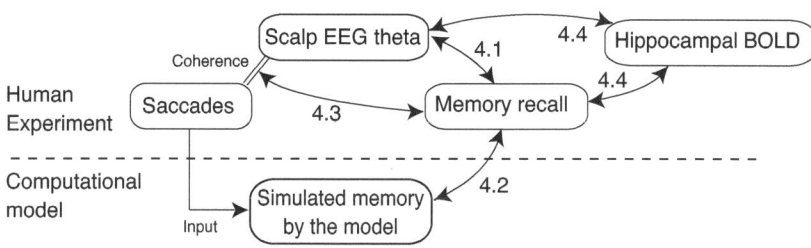

Fig. 6. Combined approach between computational model and human experimental data analysis. Major components of the computational model, eye movement, EEG theta and memory performance are systematically evaluated. Numbers of arrows indicate section number that is associated with particular experimental data analysis.

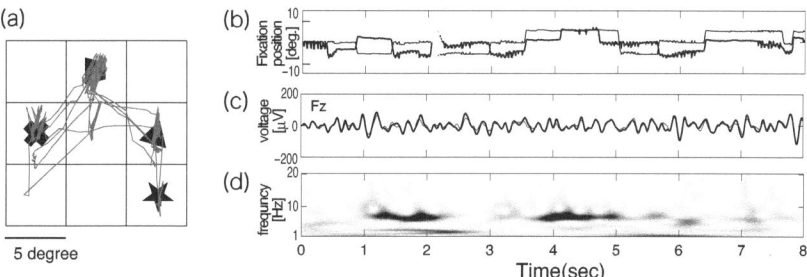

Fig. 7. An object-place memory task. (a) Visual stimulus and eye movement during 8-sec encoding period. (b) Horizontal and vertical eye movement. (c) EEG signal at the midline frontal electrode (Fz). (d) Time-frequency energy with wavelet transformation.

by a correction method consisting of EEG-EOG regression and EOG subtraction [40].

In the result, ocular artifacts are successfully removed from EEG signals (Fig. 7b). A significant increase of EEG power was found at a theta range of 6.5-7.5 Hz, in a widely distributed area from the frontal to parietal regions, while EOG signals did not show significant change in the theta band. The EEG coherence analysis also showed the EEG theta coherence increase during encoding in relation to subsequent successful recall. The location-frequency relationship in the theta band coherence corresponded well with the subsequent memory effect found in EEG power. These results are in good agreement with the prediction of the computational model. Beyond the prediction, the distributed EEG theta is considered to consist of the following three components: Frontal EEG theta of item encoding [41], parietal EEG theta of visuo-spatial control [42] and central EEG theta associated with implicit memory encoding that is likely to link with the hippocampal theta [43]. In section 4.4, we directly evaluate the relationship between the central EEG theta and hippocampal BOLD signal.

4.2   Simulated Memory Based on the Computational Model

The combined analysis between computational theory and human experimental data analysis is known as an effective approach to evaluate the specific neural dynamics associated with brain function (e.g., a reinforcement leaning in a gambing task [44]). Here we introduce a combined analysis to solve the main question, whether the neural encoding of the theta phase coding and the resultant hierarchical network of object-place memory really exist in the human brain [45]. In our analysis, human eye movement data during encoding of object-place associations were introduced to the computational model of the object-place memory based on the theta phase coding. If the theta phase precession dynamics exist in the human brain, the resultant network should correlate with human subsequent memory recall, otherwise the model is rejected.

One-time introduction of an 8-sec human eye movement data was enough to form an object-place hierarchical network in the computational model. The resultant network was evaluated by a computational recall procedure and found that an initial activation started a sequential activation representing multiple object-place associations. Importantly, both the degree of the hierarchical structure and the computational recall performance were significantly correlated with human recall performance, while simple behavioral parameters (i.e., blink frequency and fixation duration) were not significantly correlated with human recall. This indicates that the model has the ability to extract memory dependent information from eye movement data and can predict the subsequent recall performance of the participants. The current result could not reject other possible neural mechanisms, while the results are still important for indicating the relevance and functional advantage of theta phase precession dynamics in the human brain.

4.3   Eye Movement-EEG Coherence

In rats, the synchronization between sniffing behavior and hippocampal EEG is expected to have an important role in odor encoding [46]. Analogous to this evidence, the computational model of the object-place memory [34] suggests a synchronization between eye saccades and EEG theta power during successful encoding. In the following analysis, we evaluated this prediction by using a coherence value between saccade rate and EEG power [47] where the coherence value is defined to be independent of either average EEG power or average saccade rate.

The coherence value between saccade rate and frontal EEG theta power (5.0–6.5 Hz) was found to significantly increased in relation to subsequently successful recall. This result is in good agreement with the prediction. Possible concerns of ocular artifacts would be rejected by a saccade rate-EOG coherence value that was not significantly correlated with subsequent recall. An additional coherence analysis during eye fixation also demonstrated that the saccade rate-EEG theta coherence was not due to the contamination of ocular artifacts in EEG signals. These results suggest that frontal EEG theta regulates eye saccade generation during object-place memory encoding.

What is the functional role of the saccade rate–EEG theta coherence? Our hypothesis is that the intermittent increase of EEG theta power produces a set of segmented visual input sequence specially relating to object-place memory encoding. We evaluated the hypothesis by using an additional combined analysis; EEG theta power-segmented input sequences were introduced to the computational network, and the resultant computational networks were evaluated in relation to subsequent human recall. Results demonstrated that computational recalls were again significantly correlated to human recall, and its correlation was further increased in compared with the previous results without input segmentation. This result suggests that the intermittent increase of hippocampal EEG theta power contributes the memory encoding of visual input sequence. This is also in line with the proposal by Ulanovsky and Moss [48] based on the bat evidence where the intermittent hippocampal EEG theta was increased with the echolocation calls.

4.4   Simultaneous EEG-fMRI

The above results showed the important relationship among saccades, scalp EEG theta and memory performance, but it is still unclear whether the scalp EEG theta is really associated with hippocampal activity. To answer this question, we performed a simultaneous EEG-fMRI recording during the object-place memory task [49]. Simultaneous EEG-fMRI measurement is a recent technique that can demonstrate a direct relationship between scalp EEG power and BOLD signals [50]

BOLD images were acquired with a 3-T MRI scanner while performing the task, and scalp EEG were simultaneously recorded. In EEG data analysis, artifacts of MR scanning, ballistocardiogram and eye movements were sequentially corrected and translated to wavelet power. In fMRI data analysis, the difference between successful and failed encoding in relation to subsequent recall was evaluated. The subsequent recall related BOLD signals were finally compared with an expected BOLD signal calculated by EEG theta power.

In EEG results, we successfully detected the EEG theta power increase in relation to subsequently successful recall. In fMRI results, right inferior frontal regions and middle occipital region were positively correlated to subsequent recall, while the hippocampal BOLD response was negatively correlated to the subsequent recall. Both positive and negative BOLD responses are considered to be critical for successful encoding [51]. Then, we calculated a temporal correlation between the hippocampal BOLD signal and the scalp EEG theta power. It was found that the hippocampal BOLD signal was negatively correlated to the EEG theta power with 8-sec delay. This result clearly indicates that the scalp EEG theta during object-place encoding is associated with the hippocampal activity, as hypothesized in the above combined analyses. In addition to this result, the inferior frontal gyrus and the middle occipital gyrus were also found to be negatively correlated to EEG theta power. This result further indicates that theta dynamics in the cortico-hippocampal system facilitates object-place memory encoding.

5   Summary

Computational experiments showed that hippocampal theta phase coding has the ability to store episodic memory consisting of complex memory contents, such as object-place memory (Section 3). By using a computational model-human experimental data combined analysis, the model was systematically evaluated in terms of human brain activity. Table 1 shows a summary of the results of the combined analyses. All results on the scalp EEG theta, the model-based simulated memory, the saccade rate-EEG theta coherence and the EEG theta-hippocampal BOLD relationship were in good agreement with the prediction of the computational model. These results strongly support that theta phase coding in the hippocampus would play an important role in the human memory formation, as observed in the rat hippocampus.

Table 1. Summary of the combined analyses

Subsequent recall	Scalp EEG θ	Computational memory	Saccade-EEG θ coherence	Hippocampal BOLD
Successful	High	Hierahcical	Coherent	Low
Failed	Low	Random	Independent	High

For the integrative understanding of the neural dynamics and the cognitive function, it is essential for including both the proposal of the computational model and its evaluation by using the combined analysis, while it is also important to pay attention for 'epiphenomena'. Here we showed a combined approach of the object-place memory, but there is still the possibility of 'true' neural dynamics that would have better computational ability and produce a consistent understanding of experimental data. To minimize this problem, the functional advantage of the computational model should validate the model. We would further improve the functional advantage of our model by combining it with experimental evidences.

In the current study, we focused on the dynamics of memory encoding, while recall dynamics that would be characterized by the hierarchical network (Section 3.2) still remains unevaluated. According to scalp EEG evidence [52], the dynamics of theta synchronization is again expected to play an important role during episodic recall. We believe that the combined analysis will play an important role in understanding the recall dynamics implemented by cortico-hippocampal cooperation through theta synchronization.

Acknowledgments

I would like to thank Dr Yoko Yamaguchi for her helpful comments and suggestions.

References

1. Grillner, S.: The motor infrastructure: from ion channels to neuronal networks. Nat. Rev. Neurosci. 4(7), 573–586 (2003)
2. Engel, A.K., Konig, P., Kreiter, A.K., Schillen, T.B., Singer, W.: Temporal coding in the visual cortex: new vistas on integration in the nervous system. Trends Neurosci. 15(6), 218–226 (1992)
3. Squire, L.R.: Memory and the hippocampus: a synthesis from findings with rats, monkeys, and humans. Psychol. Rev. 99(2), 195–231 (1992)
4. Marr, D.: Simple memory: A theory for archicortex. Philos. Trans. R. Soc. Lond. B. Biol. Sci. 262, 23–81 (1971)
5. van Hoesen, G.W.: The parahippocampal gyrus. New observations regarding its cortical connections in the monkey. Trends Neurosci. 5, 345–350 (1982)
6. Scoville, W.B., Milner, B.: Loss of recent memory after bilateral hippocampal lesions. J. Neurol. Neurosurg. Psychiatry. 20, 11–21 (1957)
7. Tulving, E.: Elements of episodic memory. Clarendon Press, Oxford (1983)
8. Bland, B.H.: The physiology and pharmacology of hippocampal formation theta rhythms. Progress in Neurobiology 26, 1–54 (1986)
9. Ekstrom, A.D., Caplan, J.B., Ho, E., Shattuck, K., Fried, I., Kahana, M.J.: Human Hippocampal Theta Activity During Virtual Navigation. Hippocampus 15, 881–889 (2005)
10. Kay, L.M.: Theta oscillations and sensorimotor performance. PNAS 102(10), 3863–3868 (2005)
11. FitzHugh, R.: Impulses and physiological states in theoretical models of nerve membrane. Biophysical Journal 1, 445–466 (1961)
12. Hoppensteadt, F.C.: An introduction to the mathematics of neurons. Cambridge University Press, Cambridge (1986)
13. O'Keefe, J., Recce, M.L.: Phase relationship between hippocampal place units and the EEG theta rhythm. Hippocampus 3, 317–330 (1993)
14. Skaggs, W.E., McNaughton, B.L., Wilson, M.A., Barnes, C.A.: Theta phase precession in hippocampal neuronal populations and the compression of temporal sequences. Hippocampus 6, 149–172 (1996)
15. Burgess, N., Barry, C., O'Keefe, J.: An oscillatory interference model of grid cell firing. Hippocampus 17(9), 801–812 (2007)
16. Tsodyks, M.V., Skaggs, W.E., Sejnowski, T.J., McNaughton, B.L.: Population dynamics and theta rhythm phase precession of hippocampal place cell firing: A spiking neuron model. Hippocampus 6, 271–280 (1996)
17. Jensen, O., Lisman, J.E.: Hippocampal CA3 region predicts memory sequence: Accounting for the phase precession of place cells. Learning and Memory 3, 279–287 (1996)
18. Wallenstein, G.V., Hasselmo, M.E.: GABAergic modulation of hippocampal population activity: Sequence learning, place field development, and the phase precession effect. Journal of Neurophysiology 78, 393–408 (1997)
19. Mehta, M.R., Lee, A.K., Wilson, M.A.: Role of experience and oscillations in transforming a rate code into a temporal code. Nature 417, 741–746 (2002)
20. Harris, K.D., Henze, D.A., Hirase, H., Leinekugel, X., Dragoi, G., Czuko, A., Buzsaki, G.: Spike train dynamics predicts theta-related phase precession in hippocampal pyramidal cells. Nature 417, 738–741 (2002)
21. Yamaguchi, Y.: A theory of hippocampal memory based on theta phase precession. Biol. Cybern. 89, 1–9 (2003)

22. Sato, N., Yamaguchi, Y.: Memory encoding by theta phase precession in the hippocampal network. Neural Comput. 15(10), 2379–2397 (2003)
23. Wu, Z., Yamaguchi, Y.: Input-dependent learning rule for the memory of spatiotemporal sequences in hippocampal network with theta phase precession. Biol. Cybern. 90(2), 113–124 (2004)
24. Wagatsuma, H., Yamaguchi, Y.: Cognitive map formation through sequence encoding by theta phase precession. Neural Comput. 16(12), 2665–2697 (2004)
25. Hafting, T., Fyhn, M.H., Moser, M., Moser, E.I.: Phase precession and phase locking in entorhinal gridc ells. Society for Neuroscience Abstracts, Program No. 68.8 (2006)
26. Smith, M.L., Milner, B.: The role of the right hippocampus in the recall of spatial location. Neuropsychologia 19, 781–793 (1981)
27. Stepankova, K., Fenton, A.A., Pastalkova, E., Kalina, M., Bohbot, V.D.: Object-location impairment in patient with thermal lesions to the right or left hippocampus. Neuropsychologia 42, 1017–1028 (2004)
28. Gaffan, D.: Scene-specific memory for objects: a model of episodic memory impairment in monkeys with fornix transection. J. Cognit. Neurosci. 6, 305–320 (1994)
29. Eacott, M.J., Norman, G.: Integrated memory for object, place, and context in rats: a possible model of episodic-like memory? J. Neurosci. 24, 1948–1953 (2004)
30. Suzuki, W.A., Amaral, D.G.: The perirhinal and parahippocampal cortices of the Macaque monkey: cortical afferents. J. Comp. Neurol. 350, 179–205 (1994)
31. Epstein, R., Kanwisher, N.: A cortical representation of the local visual environment. Nature 392, 598–601 (1998)
32. Rolls, E.T.: Spatial view cells and the representation of place in the primate hippocampus. Hippocampus 9, 467–480 (1999)
33. de Araujo, I.E., Rolls, E.T., Stringer, S.M.: A view model which accounts for the spatial fields of hippocampal primate spatial view cells and rat place cells. Hippocampus 11(6), 699–706 (2001)
34. Sato, N., Yamaguchi, Y.: On-line formation of a hierarchical cognitive map for object-place association by theta phase coding. Hippocampus 15, 963–978 (2005)
35. Rolls, E.T., Stringer, S.M., Trappenberg, T.P.: A unified model of spatial and episodic memory. Proc. R. Soc. Lond. B 269, 1087–1093 (2002)
36. Sato, N., Yamaguchi, Y.: A hierarchical representation of object-place memory obtained by theta phase coding in the hippocampus: A computational study, Program No. 581.2, Society for Neuroscience Abstracts (2005)
37. Stevens, A., Coupe, P.: Distortion in judged spatial relations. Cognitive Psychology 10, 422–437 (1978)
38. McNamara, T.P., Hardy, J.K., Hirtle, S.C.: Subjective hierarchies in spatial memory. Journal of Experimental Psychology: Learning, Memory, and Cognition 15(2), 211–227 (1989)
39. Sato, N., Yamaguchi, Y.: Theta synchronization networks emerge during human object-place memory encoding. NeuroReport 18(5), 419–424 (2007)
40. Croft, R.J., Barry, R.J.: Removal of ocular artifact from EEG: a review. Neurophysical Clin. 30, 5–19 (2000)
41. Summerfield, C., Mangels, J.: Coherent theta-band EEG activity predicts item-context binding during encoding. NeuroImage 24, 692–703 (2005)
42. Sauseng, P., Klimesch, W., Schabus, M., Doppelmayr, M.: Fronto-parietal EEG coherence in theta and upper alpha reflect central executive functions of working memory. Int. J. Psychophysiol. 57(2), 97–103 (2005)

43. Klimesch, W., Doppelmayr, M., Russegger, H., Pachinger, T.: Theta band power in the human scalp EEG and the encoding of new information. Neuroreport 7(7), 1235–1240 (1996)
44. Tanaka, S.C., Doya, K., Okada, G., Ueda, K., Okamoto, Y., Yamawaki, S.: Prediction of immediate and future rewards differentially recruits cortico-basal ganglia loops. Nature Neurosci. 7(8), 887–893 (2004)
45. Sato, N., Yamaguchi, Y.: An evidence of a hierarchical representation of object-place memory based on theta phase coding: A computational model- human experiment combined analysis, Program No. 366.25, Society for Neuroscience Abstracts (2006)
46. Kepecs, A., Uchida, N., Mainen, Z.F.: The Sniff as a Unit of Olfactory Processing. Chemical Senses 31(2), 167–179 (2006)
47. Sato, N., Yamaguchi, Y.: EEG theta regulates eye saccade generation during human object-place memory encoding. In: Proc. the 1st international conference on cognitive neurodynamics & the 3rd shanghai international conference on physiological biophysics-cognitive neurodynamics (ICCN 2007 & SICPB 2007), Shanghai, China (2007)
48. Ulanovsky, N., Moss, C.: Hippocampal cellular and network activity in freely moving echolocating bats. Nat. Neurosci. 10(2), 224–233 (2007)
49. Sato, N., Ozaki, J.T., Someya, Y., Anami, K., Ogawa, S., Mizuhara, H., Yamaguchi, Y.: Subsequent memory dependent EEG theta correlates with hippocampal BOLD response in human. Program No. 667.1, Society for Neuroscience Abstracts (2007)
50. Mizuhara, H., Wang, L.Q., Kobayashi, K., Yamaguchi, Y.: A long-range cortical network emerging with theta oscillation in a mental task. Neuroreport 15(8), 1233–1238 (2004)
51. Daselaar, S.M., Prince, S.E., Cabeza, R.: When less means more: deactivations during encoding that predict subsequent memory. NeuroImage 23, 921–927 (2004)
52. Klimesch, W., Doppelmayr, M., Stadler, W., Pollhuber, D., Sauseng, P., Rohm, D.: Episodic retrieval is reflected by a process specific increase in human electroencephalographic theta activity. Neurosci. Lett. 302(1), 49–52 (2001)

Mechanisms for Memory-Guided Behavior Involving Persistent Firing and Theta Rhythm Oscillations in the Entorhinal Cortex

Michael E. Hasselmo, Lisa M. Giocomo, Mark P. Brandon, and Motoharu Yoshida

Center for Memory and Brain and
Program in Neuroscience, Boston University,
2 Cummington St., Boston, Massachusetts 02215,
(617) 353-1397, Fax.: (617) 358-3296
hasselmo@bu.edu

Abstract. Interactions of hippocampal and parahippocampal regions are important for memory-guided behavior. Understanding the role of these structures requires understanding the interaction of populations of neurons, including the cellular properties of neurons in structures such as the entorhinal cortex. Recent data and modeling suggest an important role for cellular mechanisms of persistent spiking and membrane potential oscillations in medial entorhinal cortex, both in mechanisms for spatial navigation and for episodic memory function. Both persistent firing and membrane potential oscillations may provide mechanisms for path integration at a cellular level based on speed-modulated head direction as a velocity signal. This path integration process thereby provides a potential mechanism for grid cell firing properties in medial entorhinal cortex. Incorporation of these processes into a larger scale model allows simulation of mechanisms for sequence encoding and episodic memory.

Keywords: episodic memory, persistent spiking, membrane potential oscillations, grid cells, theta rhythm, neuromodulation, grid cells, stellate cells, spatial navigation, path integration.

1 Introduction

Considerable research has focused on interactions of components of the hippocampal formation in memory-guided behavior at the level of interacting brain regions. This includes data showing impairments of memory-guided behavior with lesions of the hippocampus [1-8] as well as parahippocampal regions such as the entorhinal [9-12] and perirhinal cortex [13]. Understanding the role of these structures requires understanding the interaction of populations of neurons in these structures, including the cellular properties of neurons in structures such as the entorhinal cortex. This chapter will review some recent data and modeling suggesting an important role for cellular mechanisms of persistent spiking and membrane potential oscillations in medial entorhinal cortex, both in mechanisms for spatial navigation and for episodic memory function.

M. Marinaro, S. Scarpetta, and Y. Yamaguchi (Eds.): Dynamic Brain, LNCS 5286, pp. 28–37, 2008.

Electrophysiological recording in awake, behaving rats provides a link between cellular properties of individual cortical regions and the behavioral role of these neurons in behavior. In particular, the recent discovery of grid cells in the medial entorhinal cortex provides an exciting set of data relevant to the function of these circuits. Grid cells are single neurons that respond in multiple locations in the environment in a hexagonal array [14-17]. They can be characterized by their spatial frequency, their orientation and their spatial phase. The regular firing properties of grid cells suggests that neural circuits can represent stimuli along continuous dimensions of space and time.

The physiological properties of these structures may contribute to the role of the hippocampal formation in memory for the sequential order of stimuli. Hippocampal lesions impair the ability of a rat to complete specific sequences of odor stimuli [1] or to respond on the basis of the order of odor stimuli [2,3]. In addition, hippocampal lesions impair the ability to perform tasks requiring retrieval of sequences of spatial responses to different locations [4-7].

Many previous models of hippocampal function have focused on its role in spatial memory by forming representations for goal directed spatial behavior [18-21]. These models focus on learning of a global gradient to a single goal location, and do not focus on the encoding and retrieval of specific spatial trajectories within the environment. However, many memory tasks require retrieval of specific trajectories, rather than using a global gradient toward a single goal location. For example, in tasks such as the 8-arm radial maze [22], the delayed spatial alternation task [23] and the version of the Morris water maze using a new platform location on each day [24], the encoding and retrieval of previously generated trajectories can guide correct behavior in the task. Behavior in these tasks is impaired by hippocampal lesions, supporting the potential role of the hippocampus in the selective encoding and retrieval of trajectories. It has been proposed that the learning of spatial trajectories may be a special case of a general capacity for learning sequences within the hippocampus, including sequences of events in an operant task [25].

Some previous models of the hippocampus have focused on encoding and retrieval of sequences [6,26-31]. However, most previous models primarily focus on encoding associations between sequential states (items or locations). The data and modeling reviewed here takes a different approach, in which each individual state (location or sensory input) is associated with the subsequent action (movement) leading to the next state. This allows an explicit representation of the time duration of sequences during retrieval, and allows retrieval to function across continuous dimensions of space and time.

2   Cellular Mechanisms for Path Integration

Medial entorhinal cortical neurons display cellular properties that could provide the basis for path integration. Path integration essentially requires integration of a velocity signal during movement in the environment. This integration could be mediated by either graded persistent spiking or by membrane potential oscillations.

In brain slice preparations of medial entorhinal cortex, pyramidal neurons show persistent spiking activity after a depolarizing current injection or a period of repetitive synaptic input [32-35]. Pyramidal neurons in layer II of medial entorhinal cortex show persistent spiking that tends to turn on and off over periods of many seconds [32]. Pyramidal neurons in deep layers of medial entorhinal cortex can maintain spiking at different graded frequencies for many minutes [34] as shown in Figure 1. The persistent spiking appears to be due to muscarinic or metabotropic glutamate activation of a calcium-sensitive non-specific cation current [35-37]. Persistent firing has also been shown in layer III of lateral entorhinal cortex [33]. The graded persistent firing shown in deep layers form the basis for path integration. The graded persistent spiking could allow these neurons to integrate synaptic input over extended periods, causing a graded change proportional to the summed magnitude of excitatory input. If the excitatory input contains a velocity signal, it would allow path integration. The velocity signal could be provided by cells that have been shown to respond on the basis of head direction of the rat. These cells are in areas that also contain neurons responding on the basis of translational velocity. Combination of this activity could provide a speed-modulated head direction signal that would correspond to a velocity signal appropriate for path integration.

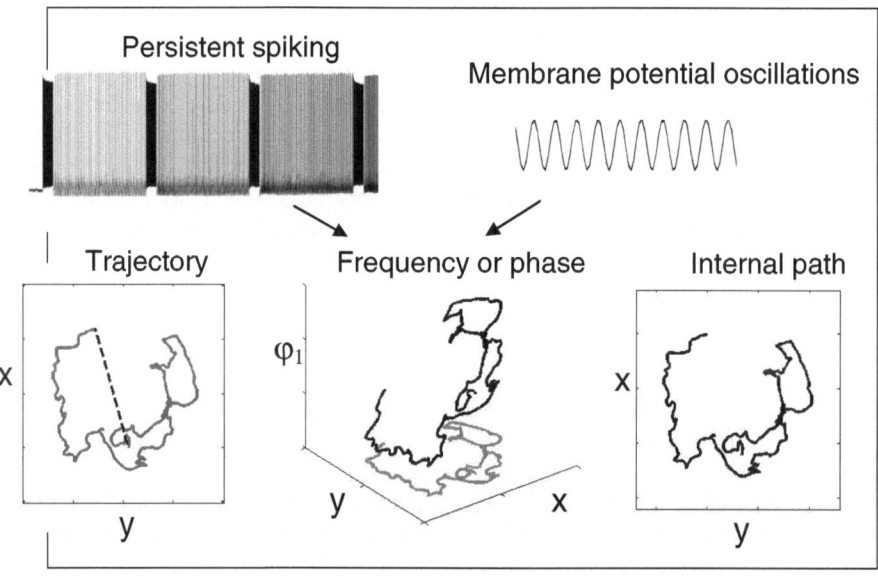

Fig. 1. Path integration could be mediated by graded persistent spiking (upper left) or membrane potential oscillations (upper right). As the rat follows a trajectory through space (left), the activation of speed-modulated head direction cells can progressively alter graded persistent firing or oscillation phase, allowing change based on the integral of movement relative to one head direction (center). The inverse transform of this internal integration representation recreates the trajectory (right).

Entorhinal neurons could also integrate excitatory input through a smooth shift in the phase of membrane potential oscillations in neurons that show intrinsic membrane potential oscillations. Entorhinal layer II stellate cells show subthreshold membrane potential oscillations when depolarized near firing threshold [38,39]. These are small oscillations of a few millivolts in amplitude that can influence the timing of action potentials [40] and can contribute to network oscillations [41,42]. The oscillations appear to be due to a hyperpolarization activated cation current or h-current [43], that differs in time constant along the dorsal to ventral axis [44]. Depolarizing input increases the frequency of these oscillations such that the phase of the oscillation integrates the depolarizing input over time. Membrane potential oscillations do not usually appear in layer II or layer III pyramidal cells [39], but are observed in layer V pyramidal cells, where they may be caused by M-current [45]. Membrane potential oscillations do not appear in neurons of the lateral entorhinal cortex [46]. The frequency of membrane potential oscillations differs systematically along the dorsal to ventral axis of the medial entorhinal cortex [47]. Modeling shows how voltage-dependent modulation of the frequency of these oscillations could underlie differences in grid cell firing properties along the dorsal to ventral axis [47-50]. Excitatory input from head direction cells can alter the frequency of these oscillations, thereby shifting the relative phase of individual oscillations. The shift in phase is proportional to the integral of prior input, allowing phase to provide an alternate mechanism for integration of a velocity signal.

3 Possible Mechanisms of Grid Cell Firing

The regular spatial firing of grid cells in entorhinal cortex indicates that these neurons have accurate information about spatial location. This could be obtained by a number of different physiological mechanisms for path integration. Thus, there are a number of different ways in which grid cells firing properties could be modeled. These include two broad categories of models dependent on either: 1. network attractor dynamics, or 2. intrinsic neuron properties. These two broad classes of models are compatible. Both mechanisms could exist in parallel.

Previous published models that have simulated grid cell firing properties based on attractor dynamics within a network of neurons [17,51]. The grid cell pattern arises from circularly symmetric connectivity with nearby excitatory connections and longer-range inhibitory connections, giving a "Mexican-hat" profile of connectivity strength. Different mechanisms are used for training this connectivity pattern in different models. An alternate network model utilizes neurons that fire with a high spatial frequency in the environment. If these cells with different high spatial frequencies interact, it causes a pattern of interference resulting in the lower frequency pattern corresponding to the grid cell firing properties [52]. These models account for certain aspects of grid cell firing, but do not yet describe how the high spatial frequency firing can be obtained.

Within the class of intrinsic neuron properties, there are multiple related mechanisms that could underlie grid cell firing. These could depend on either membrane potential oscillations or persistent firing. The first single cell model of grid cell firing used interference of membrane potential oscillations to obtain grid cell firing properties [49,50,53]. In this model, excitatory synaptic input from speed-modulated head direction cells alters the frequency of membrane potential oscillations within single dendrites, thereby shifting the phase of firing relative to other dendrites.

Experimental data from entorhinal cortex demonstrates how intracellular membrane potential oscillations could provide this mechanism of the model [47,48]. In particular, the model depends upon the fact that entorhinal stellate cells show membrane potential oscillations around theta frequency [38,39,43,47], and these membrane potential oscillations are increased in frequency by depolarization. There are excitatory projections to entorhinal cortex stellate cells from the postsubiculum [54] and deep layers of entorhinal cortex [16], both of which contain head direction cells. Thus, active head direction cells can directly depolarize entorhinal stellate cells. The Burgess model [49] uses the speed modulated head direction $\bar{h}(t)$ to regulate the intrinsic oscillation frequency of different dendritic branches of entorhinal stellate cells as summarized by the following equation:

$$g(t) = \Theta[\prod_i \{\cos(f\,2\pi t) + \cos(f\,2\pi(t + B\int_0^t \bar{h}_i(\tau)d\tau) + \vec{\varphi}_0)\}] \tag{1}$$

Where $g(t)$ represents the firing of a single grid cell over time. Π_i represents the product of the different dendritic oscillations receiving input from different components of the head direction vector with index i, and Θ represents a Heaviside step function output (the model has output 1 for any value above a threshold). The intrinsic oscillations have frequency f, and the modulation of this frequency is scaled by a constant B resulting in the beat frequency $f*B=f_b$. The dendritic branches each have an initial phase φ_0. This creates grid cell firing fields with spacing dependent upon modulation of the intrinsic oscillation frequency f_b [47,48]. The sum of the somatic oscillation with frequency f and each dendritic oscillation with frequency $f + fB\bar{h}(t)$ results in an interference pattern that has an envelope with a frequency equal to the difference of the two frequencies $fB\bar{h}(t)$.

The model described above makes a number of assumptions that are strong predictions about network properties, and are not completely consistent with available data. This includes: 1.) The model requires oscillations of distinct frequencies in different parts of one neuron, whereas simulations of stellate cells suggest strong synchronization properties for oscillations in single cells. 2.) The model breaks down with noisy oscillations, whereas data suggests considerable noise in oscillation phase. 3.) The model requires intrinsic oscillations for formation of grid cells, whereas intracellular recording shows an absence of membrane potential oscillations in layer III cells in entorhinal cortex despite the fact that they show grid cell properties . 4.) The model does not yet address the formation of grid cells in deep layers that show firing strongly dependent on head direction.

The model also makes basic assumptions about head direction cells that do not necessarily fit the data. 5.) The model requires input from head direction cells with 60 degree differences in angle of selectivity. 6.) The model uses cosine tuned head direction cells, whereas most experimentally described head direction cells have narrower tuning functions with a triangular shape. 7.) The model requires speed-modulated head direction cells, whereas most actual head direction cells are not speed-modulated. 8.) The simplest model uses head direction cells that provide both positive and negative inputs. To match real head direction cells, these inputs can be rectified to be all positive, but this requires separate excitatory and inhibitory inputs from coupled pairs of head direction cells at 180 degree angle differences.

Other models of grid cells can be created based on intrinsic properties that integrate the speed-modulated head direction signal. The interference of oscillations could take place in different ways. For example, the oscillations could cause dendritic spiking that then interacts to cause somatic spiking. Alternately, oscillations in individual entorhinal stellate cells could be modulated by pairs of head direction inputs that shift the frequency up or down. If these cells spike on the peak of each oscillation, they will not show any location modulation in firing rate, but neurons could show grid cell responding if they receive convergent synaptic input from cells whose phase is determined by head direction inputs with 60 degree differences.

Other models can utilize persistent spiking properties of neurons. For example, instead of spiking dependent upon membrane potential oscillations, rhythmic spiking could be provided by intrinsic graded persistent spiking. If the frequency of this spiking is increased and decreased by head direction input, it can cause spiking with phase dependent upon location. Other neurons receiving input from these graded persistent spiking neurons can fire in a grid cell pattern dependent on synchronous input from neurons receiving different head direction inputs tuned at 60 degree intervals. Alternately, the grid cell firing could depend upon the intrinsic tendency for the persistent firing to turn off and on during sustained depolarization. This could allow a neuron receiving opposite head direction input to show spatially periodic firing dependent upon the integration of the opposite head direction inputs.

4 Circuit Model of Episodic Memory

The interaction of head direction cells and grid cells described here provides a potential mechanism for episodic memory involving the storage of trajectories through space and time [55]. This model uses a functional loop that encodes and retrieves trajectories via three stages: 1.) head direction cells update grid cells, 2.) grid cells update place cells, and 3.) place cells activate associated head direction activity [55]. This model is consistent with the anatomical connectivity. The head direction cells could update grid cells via projections from the postsubiculum (dorsal presubiculum) to the medial entorhinal cortex [54,56,57]. Grid cells can update place cells via the extensive projections from entorhinal cortex layer II to dentate gyrus and CA3 and from layer III to region CA1 [58,59]. Place cells can become associated with head direction activity via projections from region CA1 to the subiculum [58,60], and projections from the dorsal and distal regions of the subiculum to the postsubiculum and medial entorhinal cortex [61], which contain head direction cells.

During initial encoding of a trajectory, this head direction cell activity vector is set by the actual head direction of the rat during exploration, and associations are encoded between place cell activity and head direction activity. During retrieval, the head direction activity depends upon synaptic input from place cell representations.

The model presented here is effective for performing episodic encoding and retrieval of trajectories in simulations [55], including trajectories based on experimental data or trajectories created by an algorithm replicating foraging movements of a rat in an open field [48]. During encoding, a series of place cells are created associated with particular locations. Each place cell is also associated with input from the grid cell population activity and with the head direction vector that occurred during the initial movement from that location. For retrieval, the simulation is cued with the grid cell phase vector and head direction vector present at the start location. The head direction vector updates the grid cell phase vector, which alters the activity of grid cells. The grid cell firing drives place cells associated with subsequent locations on the trajectory.

The activation of each new place cell activates a new head direction vector h_p associated with that place cell. This new head direction vector then drives the further update of dendritic phases of grid cells. This maintenance of the head direction vector might require graded persistent spiking [34] of head direction cells in deep layers of entorhinal cortex. The retrieval of the place cell activity representing the state drives the retrieval of the new head direction vector representing the action from that state. This action is then used for a period of time to update the grid cell state representation until a new place cell representation is activated.

Because retrieval of the trajectory depends on updating of phase by head direction cells, this allows retrieval of a trajectory at a time course similar to the initial encoding. This can allow effective simulation of the slow time course of place cell replay observed during REM sleep [62]. The spread of activity from place cells to cells coding head direction could contribute to patterns of firing in the postsubiculum that appear as cells responding dependent on both place and head direction [63]. These cells might code the action value for retrieval of a trajectory from a particular location, firing only when actual head direction matches the head direction previously associated with specific place cell activity. The strong theta phase specificity of these cells could be due to separate dynamics for encoding and retrieval within each cycle of theta rhythm [64]. These cells might selectively fire during the retrieval phase.

Acknowledgments. Research supported by Silvio O. Conte Center grant NIMH MH71702, NIMH R01 60013, NIMH MH61492, NIMH MH60450, NSF SLC SBE 0354378 and NIDA DA16454 (part of the CRCNS program).

References

1. Agster, K.L., Fortin, N.J., Eichenbaum, H.: The hippocampus and disambiguation of overlapping sequences. J. Neurosci. 22, 5760–5768 (2002)
2. Kesner, R.P., Gilbert, P.E., Barua, L.A.: The role of the hippocampus in memory for the temporal order of a sequence of odors. Behav. Neurosci. 116, 286–290 (2002)
3. Fortin, N.J., Agster, K.L., Eichenbaum, H.B.: Critical role of the hippocampus in memory for sequences of events. Nature Neuroscience 5, 458–462 (2002)
4. Kesner, R.P., Novak, J.M.: Serial position curve in rats: role of the dorsal hippocampus. Science 218, 173–175 (1982)
5. M'Harzi, M., Palacios, A., Monmaur, P., Willig, F., Houcine, O., Delacour, J.: Effects of selective lesions of fimbria-fornix on learning set in the rat. Physiol. Behav. 40, 181–188 (1987)
6. Hasselmo, M.E., Eichenbaum, H.: Hippocampal mechanisms for the context-dependent retrieval of episodes. Neural Netw. 18, 1172–1190 (2005)
7. Lee, I., Jerman, T.S., Kesner, R.P.: Disruption of delayed memory for a sequence of spatial locations following CA1- or CA3-lesions of the dorsal hippocampus. Neurobiol. Learn. Mem. 84, 138–147 (2005)

8. Eichenbaum, H., Cohen, N.J.: From conditioning to conscious recollection: Memory systems of the brain. Oxford University Press, New York (2001)
9. Steffenach, H.A., Witter, M., Moser, M.B., Moser, E.I.: Spatial memory in the rat requires the dorsolateral band of the entorhinal cortex. Neuron 45, 301–313 (2005)
10. Bunsey, M., Eichenbaum, H.: Critical role of the parahippocampal region for paired-associate learning in rats. Behavioral Neuroscience 107, 740–747 (1993)
11. McGaughy, J., Koene, R.A., Eichenbaum, H., Hasselmo, M.E.: Cholinergic deafferentation of the entorhinal cortex in rats impairs encoding of novel but not familiar stimuli in a delayed nonmatch-to-sample task. J. Neurosci. 25, 10273–10281 (2005)
12. Buckmaster, C.A., Eichenbaum, H., Amaral, D.G., Suzuki, W.A., Rapp, P.R.: Entorhinal cortex lesions disrupt the relational organization of memory in monkeys. J. Neurosci. 24, 9811–9825 (2004)
13. Zola-Morgan, S., Squire, L.R., Amaral, D.G., Suzuki, W.A.: Lesions of perirhinal and parahippocampal cortex that spare the amygdala and hippocampal formation produce severe memory impairment. J. Neurosci. 9, 4255–4270 (1989)
14. Fyhn, M., Hafting, T., Treves, A., Moser, M.B., Moser, E.I.: Hippocampal remapping and grid realignment in entorhinal cortex. Nature 446, 190–194 (2007)
15. Hafting, T., Fyhn, M., Molden, S., Moser, M.B., Moser, E.I.: Microstructure of a spatial map in the entorhinal cortex. Nature 436, 801–806 (2005)
16. Sargolini, F., Fyhn, M., Hafting, T., McNaughton, B.L., Witter, M.P., Moser, M.B., Moser, E.I.: Conjunctive representation of position, direction, and velocity in entorhinal cortex. Science 312, 758–762 (2006)
17. Fuhs, M.C., Touretzky, D.S.: A spin glass model of path integration in rat medial entorhinal cortex. J. Neurosci. 26, 4266–4276 (2006)
18. Trullier, O., Meyer, J.A.: Animat navigation using a cognitive graph. Biol. Cybern. 83, 271–285 (2000)
19. Burgess, N., Donnett, J.G., Jeffery, K.J., O'Keefe, J.: Robotic and neuronal simulation of the hippocampus and rat navigation. Philos. Trans. R. Soc. Lond. B Biol. Sci. 352, 1535–1543 (1997)
20. Foster, D.J., Morris, R.G., Dayan, P.: A model of hippocampally dependent navigation, using the temporal difference learning rule. Hippocampus 10, 1–16 (2000)
21. Touretzky, D.S., Redish, A.D.: Theory of rodent navigation based on interacting representations of space. Hippocampus 6, 247–270 (1996)
22. Bunce, J.G., Sabolek, H.R., Chrobak, J.J.: Intraseptal infusion of the cholinergic aganist carbachol impairs delayed-non-match-to-sample radial arm maze performance in the rat. Hippocampus 14, 450–459 (2004)
23. Ennaceur, A., Neave, N., Aggleton, J.P.: Neurotoxic lesions of the perirhinal cortex do not mimic the behavioural effects of fornix transection in the rat. Behav. Brain Res. 80, 9–25 (1996)
24. Buresova, O., Bolhuis, J.J., Bures, J.: Differential effects of cholinergic blockade on performance of rats in the water tank navigation task and in a radial water maze. Behav. Neuro-sci. 100, 476–482 (1986)
25. Eichenbaum, H., Dudchenko, P., Wood, E., Shapiro, M., Tanila, H.: The hippocampus, memory, and place cells: is it spatial memory or a memory space? Neuron 23, 209–226 (1999)
26. Redish, A.D., Touretzky, D.S.: The role of the hippocampus in solving the Morris water maze. Neural Comput. 10, 73–111 (1998)
27. Jensen, O., Lisman, J.E.: Hippocampal CA3 region predicts memory sequences: accounting for the phase precession of place cells. Learning and Memory 3, 279–287 (1996)

28. Levy, W.B.: A sequence predicting CA3 is a flexible associator that learns and uses context to solve hippocampal-like tasks. Hippocampus 6, 579–590 (1996)
29. Wallenstein, G.V., Hasselmo, M.E.: GABAergic modulation of hippocampal population activity: sequence learning, place field development, and the phase precession effect. J. Neurophysiol. 78, 393–408 (1997)
30. Jensen, O., Lisman, J.E.: Theta/gamma networks with slow NMDA channels learn sequences and encode episodic memory: role of NMDA channels in recall. Learning and Memory 3, 264–278 (1996)
31. Zilli, E.A., Hasselmo, M.E.: Modeling the role of working memory and episodic memory in behavioral tasks. Hippocampus 18(2), 193–209 (2008)
32. Klink, R., Alonso, A.: Muscarinic modulation of the oscillatory and repetitive firing properties of entorhinal cortex layer II neurons. J. Neurophysiol. 77, 1813–1828 (1997)
33. Tahvildari, B., Fransen, E., Alonso, A.A., Hasselmo, M.E.: Switching between "On" and "Off" states of persistent activity in lateral entorhinal layer III neurons. Hippocampus 17, 257–263 (2007)
34. Egorov, A.V., Hamam, B.N., Fransen, E., Hasselmo, M.E., Alonso, A.A.: Graded persistent activity in entorhinal cortex neurons. Nature 420, 173–178 (2002)
35. Fransén, E., Tahvildari, B., Egorov, A.V., Hasselmo, M.E., Alonso, A.A.: Mechanism of graded persistent cellular activity of entorhinal cortex layer v neurons. Neuron 49, 735–746 (2006)
36. Shalinsky, M.H., Magistretti, J., Ma, L., Alonso, A.A.: Muscarinic activation of a cation current and associated current noise in entorhinal-cortex layer-II neurons. J. Neuro. physiol. 88, 1197–1211 (2002)
37. Yoshida, M., Fransen, E., Hasselmo, M.E.: mGlur-dependent persistent firing in entorhinal cortex layer III neurons. Eur. J. Neurosci (in press, 2008)
38. Alonso, A., Llinas, R.R.: Subthreshold Na-dependent theta-like rhythmicity in stellate cells of entorhinal cortex layer II. Nature 342, 175–177 (1989)
39. Alonso, A., Klink, R.: Differential electroresponsiveness of stellate and pyramidal-like cells of medial entorhinal cortex layer II. J. Neurophysiol. 70, 128–143 (1993)
40. Fransen, E., Alonso, A.A., Dickson, C.T., Magistretti, J., Hasselmo, M.E.: Ionic mechanisms in the generation of subthreshold oscillations and action potential clustering in entorhinal layer II stellate neurons. Hippocampus 14, 368–384 (2004)
41. Acker, C.D., Kopell, N., White, J.A.: Synchronization of strongly coupled excitatory neurons: relating network behavior to biophysics. J. Comput. Neurosci. 15, 71–90 (2003)
42. Alonso, A., Garcia-Austt, E.: Neuronal sources of theta rhythm in the entorhinal cortex of the rat. I. Laminar distribution of theta field potentials. Experimental Brain Research 67, 493–501 (1987)
43. Dickson, C.T., Magistretti, J., Shalinsky, M.H., Fransen, E., Hasselmo, M.E., Alonso, A.: Properties and role of I(h) in the pacing of subthreshold oscillations in entorhinal cortex layer II neurons. J. Neurophysiol. 83, 2562–2579 (2000)
44. Giocomo, L.M., Hasselmo, M.E.: Time constant of I(h) differs along dorsal to ventral axis of medial entorhinal cortex. Journal of Neuroscience (in press)
45. Yoshida, M., Alonso, A.: Cell-type specific modulation of intrinsic firing properties and subthreshold membrane oscillations by the m(kv7)-current in neurons of the entorhinal cortex. J. Neurophysiol. 98, 2779–2794 (2007)
46. Tahvildari, B., Alonso, A.: Morphological and electrophysiological properties of lateral entorhinal cortex layers II and III principal neurons. J Comp. Neurol. 491, 123–140 (2005)

47. Giocomo, L.M., Zilli, E.A., Fransen, E., Hasselmo, M.E.: Temporal frequency of sub-threshold oscillations scales with entorhinal grid cell field spacing. Science 315, 1719–1722 (2007)
48. Hasselmo, M.E., Giocomo, L.M., Zilli, E.A.: Grid cell firing arise from interfer-ence of theta frequency membrane potential oscillations in single neurons. Hippocampus 17, 1252–1271 (2007)
49. Burgess, N., Barry, C., O'Keefe, J.: An oscillatory interference model of grid cell firing. Hippocampus 17, 801–812 (2007)
50. Burgess, N., Barry, C., Jeffery, K.J., O'Keefe, J.: A grid and place cell model of path inte-gration utilizing phase precession versus theta, Computational Cognitive Neuroscience Meeting, Computational Cognitive Neuroscience Meeting, Washington, D.C (2005)
51. McNaughton, B.L., Battaglia, F.P., Jensen, O., Moser, E.I., Moser, M.B.: Path integration and the neural basis of the 'cognitive map'. Nat. Rev. Neurosci. 7, 663–678 (2006)
52. Blair, H.T., Welday, A.C., Zhang, K.: Scale-invariant memory representations emerge from moire interference between grid fields that produce theta oscillations: a computa-tional model. J. Neurosci. 27, 3211–3229 (2007)
53. O'Keefe, J., Burgess, N.: Dual phase and rate coding in hippocampal place cells: theoreti-cal significance and relationship to entorhinal grid cells. Hippocampus 15, 853–866 (2005)
54. Caballero-Bleda, M., Witter, M.P.: Regional and laminar organization of projections from the presubiculum and parasubiculum to the entorhinal cortex: an anterograde tracing study in the rat. J. Comp. Neurol. 328, 115–129 (1993)
55. Hasselmo, M.E.: Arc length coding by interference of theta frequency oscillations underlie context-dependent hippocampal unit data and episodic memory function. Learn Mem. 14, 782–794 (2007)
56. van Groen, T., Wyss, J.M.: The postsubicular cortex in the rat: characterization of the fourth region of the subicular cortex and its connections. Brain Res. 529, 165–177 (1990)
57. Kohler, C.: Intrinsic projections of the retrohippocampal region in the rat brain. I. The subicular complex. J. Comp. Neurol. 236, 504–522 (1985)
58. Amaral, D.G., Witter, M.P.: The 3-dimensional organization of the hippocampal formation - A review of anatomical data. Neurosci. 31, 571–591 (1989)
59. Solstad, T., Moser, E.I., Einevoll, G.T.: From grid cells to place cells: a mathematical model. Hippocampus 16, 1026–1031 (2006)
60. Swanson, L.W., Wyss, J.M., Cowan, W.M.: An autoradiographic study of the organi-zation of intrahippocampal association pathways in the rat. J. Comp. Neurol. 181, 681–716 (1978)
61. Naber, P.A., Witter, M.P.: Subicular efferents are organized mostly as parallel projec-tions: a double-labeling, retrograde-tracing study in the rat. J. Comp. Neurol. 393, 284–297 (1998)
62. Louie, K., Wilson, M.A.: Temporally structured replay of awake hippocampal ensemble activity during rapid eye movement sleep. Neuron 29, 145–156 (2001)
63. Cacucci, F., Lever, C., Wills, T.J., Burgess, N., O'Keefe, J.: Theta-modulated place-by-direction cells in the hippocampal formation in the rat. J. Neurosci. 24, 8265–8277 (2004)
64. Hasselmo, M.E., Bodelon, C., Wyble, B.P.: A proposed function for hippocampal theta rhythm: separate phases of encoding and retrieval enhance reversal of prior learning. Neu-ral Comput. 14, 793–817 (2002)

Encoding and Replay of Dynamic Attractors with Multiple Frequencies: Analysis of a STDP Based Learning Rule

Silvia Scarpetta, Masahiko Yoshioka, and Maria Marinaro

[1] Dept.of Physics "E.R.Caianiello", Univ.of Salerno
[2] IIASS, INFM and INFN Gruppo Coll. di Salerno, Italy
silvia@sa.infn.it

Abstract. In this paper we review a model of learning based on the Spike Timing Dependent Plasticity (STDP), introduced in our previous works, and we extend the analysis to the case of multiple frequencies, showing how the learning rule is able to encode multiple spatio-temporal oscillatory patterns with distributed frequencies as dynamical attractors of the network.

After learning, each encoded oscillatory spatio-temporal pattern who satisfy the stability condition forms a dynamical attractor, such that, when the state of the system falls in the basin of attraction of one such dynamical attractor, it is recovered with the same encoded phase relationship among units. Here we extend the analysis introduced in our previous work, to the case of distributed frequencies, and we study the relation between stability of multiple frequencies and the shape of the learning window. The stability of the dynamical attractors play a critical role. We show that imprinting into the network a spatio-temporal pattern with a new frequency of oscillation can destroy the stability of patterns encoded with different frequency of oscillation. The system is studied both with numerical simulations, and analytically in terms of order parameters when a finite number of dynamic attractors are encoded into the network in the thermodynamic limit.

1 Introduction

Recent advances in experimental brain research have generated renewed awareness and appreciation that the brain operates as a complex nonlinear dynamic system. From the pioneer work of V.Braitenberg E.R.Caianiello F Lauria and N. Onesto titled "A system of coupled oscillators as a functional model of neural assemblies" published in 1959 [1] the idea that brain can be modeled as a system of coupled oscillators has attracted a lot of attention, especially in recent years.

The growing interest in neuronal oscillations [3,4] is a results of several developments. Whereas in the past one could only observe oscillations in the EEG, nowadays phase locked oscillatory activity has been recorded in many in-vitro and in-vivo systems, and it's also possible to create them under controlled situations [5], [6]. Biophysical studies revealed that neural circuits and even single

M. Marinaro, S. Scarpetta, and Y. Yamaguchi (Eds.): Dynamic Brain, LNCS 5286, pp. 38–60, 2008.

neurons are able to resonate and oscillate at multiple frequencies [7], [8], These results led to the tantalizing conjecture that synchronized and phase locked oscillatory networks play a fundamental role in perception, memory, and sensory computation [3, 4, 45]. The research field of "neural oscillations" has created an interdisciplinary platform that cut across mathematics, neuroscience, biophysics, cognitive psychology, computational modeling and physics. At the same time, it have been observed in hippocampus and other brain areas, that spatio-temporal patterns of neural activity of awake state are replayed during sleep [9,10,11]. The observed replay is supposed to play a role in the process of memory consolidation [11]. Not only many repeating spikes sequences has been observed replayed compressed in time [9], but also patterns of firing rate correlations between neurons have been observed (in rat mPFC during a repetitive sequence task) to be preserved during subsequent sleep, and compressed in time [11]. Notably, reverse replay of behavioral sequences in hippocampal place cell s during the awake state has also been observed [12].

Recent experimental findings further underlined the importance of dynamics by showing that long term changes in synaptic strengths depend on the precise relative timing of pre- and post-synaptic firing ([27, 31, 32, 30, 29, 28, 33]) which suggest that precise timing of neurons activity within neuronal networks could represent information.

Many oscillatory models has been proposed for different areas of the brain, as coupled oscillators in a phase-representation (see, among others, [13, 14, 15, 16, 20, 22, 23, 43, 44] and references therein), chaotic oscillators models [21], reduced models of coupled Excitatory and Inhibitory (E-I) units [24, 25] and coupled spiking neurons [17, 18, 19], and many experimental features has been accounted by these models.

In this paper we show how a learning rule [2, 25] based on the STDP learning window is able to encode multiple periodic spatio-temporal patterns with different frequencies as attractors of the network dynamics, such that these spatio-temporal patterns are replayed spontaneously, depending from initial conditions. With respect to our previous studies, here we extend the analysis to the case of distributed frequencies, studying how the same network can be resonant to different frequencies. This is particularly interesting since experimentally has been observed that the same system is able to sustain different rhythms (see also other works in this volume). In our model, learned memory states can be encoded by both the frequency and the phases of the oscillating neural populations, enabling more efficient and robust information coding than in conventional models of associative memory.

This paper is composed of three parts. In the first part we describe the model and the learning rule. After learning, the dynamic of the network of N coupled nonlinear units is analyzed, looking for the analytical conditions under which the learned patterns are dynamical attractors of the dynamics. In the second part, we extend the order parameter analysis and the stability analysis that we introduce in [2] to the case of encoded patterns with different frequencies. We find that when we encode multiple patterns with distributed frequencies in the net, stability of each attractors depends from the set of frequencies encoded. Encoding a pattern with a new frequency can destroy the stability of patterns encoded with other frequencies. In the third part we discuss the learning rule, and we investigate here (1) the relation between frequencies encodable as stable attractors and shape of the learning window, (2) the relation between shape of learning window and time scale of retrieval replay. The case of reverse replay is also discussed.

2   The Model

The model has been introduced in [2]. We review its main ingredients here. Dynamic equations of unit x_i is

$$\tau_d \frac{d}{dt} x_i = -x_i + F(h_i) \qquad (1)$$

where transfer function F(h) denotes the nonlinear sigmoidal input-output relationship of neurons, τ_d is the time constant of unit x_i (for simplicity, assume the same for all units), and local field h_i is defined by

$$h_i = \sum_j J_{ij} x_j \qquad (2)$$

where J_{ij} is the connection after the learning procedure. Spontaneous activity dynamics of the coupled nonlinear system is determined by the function F(h) and by the coupling matrix J_{ij}. TheJ_{ij} is learned using the proposed learning procedure described in next section (2.1) We keep explicitly the time constant τ_d to see its interaction with the time scale of plasticity and its role to determine the time scale of the spontaneous dynamics after learning, i.e. the frequency of replay.

2.1   Spike-Timing-Dependent Plasticity and the Learning Rule for Dynamical Attractors

Plasticity of synaptic connections is regarded as a cellular basis for the developmental and learning-related changes in the central nervous system. In neocortical and hippocampal pyramidal cells has been found [27,31,32,30,28,33,34], that the synaptic strength increases (long-term potentiation (LTP)) or decreases (long-term depression (LTD)), whether the pre-synaptic spike precedes or follows the post-synaptic one by few ms, with a degree of change that depend from the delay between pre and post-synaptic activity via a learning window that is asymmetric with respect to time reversal. This temporally asymmetric form of synaptic plasticity was also employed at the same time or earlier in a model for auditory localization by relative timing of spikes from two ears [37, 38]. Here we analyze the behavior of the simple model of eqs. (1,2), when we use the asymmetric time-dependent learning rule, proposed in [26,25,2], inspired to the experimental findings cited above. According to the asymmetric time-dependent learning rule that we study here, the change that occurs in interval $[-T, 0]$ can be formulated as follows:

$$\delta J_{ij}(T) \propto \frac{1}{T} \int_{-T}^{0} dt \int_{-T}^{0} dt' \, x_i(t) A(t - t') x_j(t') \tag{3}$$

for synaptic weight J_{ij}, where x_j and x_i are the pre- and post-synaptic activities. The learning window $A(\tau)$ is the measure of the strength of synaptic change when there's a time delay τ between pre and post-synaptic activity.

Activity-dependent modification of synaptic strengths due to the proposed learning rule in eq. (3) is sensitive to correlations between pre- and post-synaptic firing over timescales of tens of ms when the range of $A(\tau)$ is tens of ms.

Note that eq. (3) reduces to the conventional Hebbian one (used, e.g., in [47]), $\delta C_{ij} \propto \frac{1}{T} \int_0^T dt \, x_i(t) x_j(t)$, when $A(\tau) \propto \delta(\tau)$. However, to model the experimental results of STDP such as [28, 34] the kernel $A(\tau)$ should be an asymmetric function of τ, mainly positive (LTP) for $\tau > 0$ and mainly negative (LTD) for $\tau < 0$. The shape of $A(\tau)$ strongly affect J_{ij} and the dynamics of the networks, as discussed in the following. Examples of learning windows used here are shown in Figs 3(A) and 4(A).

If the learning rule proposed here is used with a proper learning window, then the network is able to memorize multiple dynamical attractors and replay them selectively. This is studied both numerically and analytically for the periodic patterns as described in the following.

Let us present an input $I_j^\mu(t)$ to our network and we apply our learning rule (3). In the brain, cholinergic modulation can affect the strengths of long-range connections; these are apparently reduced strongly during learning [35, 36]. In our model we therefore make the assumption that connections J_{ij} are ineffective during learning (while they are plastic and learn their new value), and the network dynamics is driven by the external input $I_j^\mu(t)$, $\tau_d \dot{x}_j = -x_j + I_j^\mu(t)$, giving $x_j(t) = L_j^\mu(t)$. Let us consider the oscillatory input I_j^μ such that activity being encoded is L_j^μ ($\mu = 1..., P$, $j = 1, ..., N$)

$$L_j^\mu(t) = 1/2(1 + \cos(\omega_\mu t - \phi_j^\mu)) \tag{4}$$

characterized by amplitudes 1, frequency ω_μ, and phases ϕ_j^μ on units j. We study analytically the system when the phases ϕ_j^μ of the encoded patterns are chosen randomly in $[0, 2\pi)$. In our previous study [2] all frequencies were assumed equals $\omega_\mu = \omega_0$, while in the present study we let ω^μ to be arbitrary values. Spatio-temporal patterns L_i^μ are assumed to be positive-valued since these patterns represent spatio-temporal firing rates. We can rewrite eq. (4) as

$$L_j^\mu(t) = \frac{1}{2} + \frac{1}{4}(\xi_j^\mu e^{-i\omega_\mu t} + \text{c.c.}), \tag{5}$$

where c.c. is complex conjugate, and we denote with ξ^μ the complex vector whose components

$$\xi_j^\mu = e^{i\phi_j^\mu} \tag{6}$$

represents the phase shifts among units of the encoded oscillatory pattern μ.

We substite $x_j(t) = L_j^\mu(t)$ in eq. (3) to calculate J_{ij}, taking the limit $T \to \infty$, and assuming $A(\tau)$ decays exponencially or even faster, we obtain that the change that occurs in interval $[-\infty, 0]$ is

$$J_{ij}^\mu = \frac{1}{N}\mathrm{Re}[\tilde{A}(\omega_\mu)\, \xi_i^\mu \xi_j^{\mu*}] + \frac{2}{N}\tilde{A}(0) = \frac{1}{N}|a^\mu|\cos(\phi_i^\mu - \phi_j^\mu + \varphi_\mu) + \frac{2}{N}\tilde{A}(0) \quad (7)$$

where we have explicitly used the conventional normalization factor $1/N$ for convenience in doing the mean field calculations, while the factor

$$\tilde{A}(\omega_\mu) = a^\mu = |a^\mu|e^{i\varphi_\mu} = \int_{-\infty}^{\infty} d\tau\, A(\tau)e^{-i\omega_\mu \tau} \quad (8)$$

is the Fourier transform of the learning window. Index μ in a^μ and φ_μ refers to the dependency from the encoded frequency ω_μ of pattern μ. The factor $\tilde{A}(\omega_\mu)$ comes from eq. (3) when $x_i(t)$ are oscillatory, and it can be thought as an effective learning rate at a frequency ω_μ.

When we encode multiple patterns $\mu = 1, 2, ...P$, the learned weights are sums of contributions from individual patterns. After learning P patterns, each with frequency ω_μ and phase-shift vector ξ^μ, one get the connections

$$J_{ij} = \sum_{\mu=1}^{P} J_{ij}^\mu = \frac{1}{N}\sum_{\mu=1}^{P}\mathrm{Re}\left(\tilde{A}(\omega_\mu)\xi_i^\mu \xi_j^{\mu*}\right) + \frac{b}{N}$$

$$= \frac{1}{N}\sum_{\mu=1}^{P}|a^\mu|\cos(\phi_i^\mu - \phi_j^\mu + \varphi_\mu) + \frac{b}{N} \quad (9)$$

with

$$b = 2P\tilde{A}(0) \equiv 2P\int_{-\infty}^{\infty} A(t)dt \quad (10)$$

The dependence of the neural connections J_{ij} on $\xi_i^\mu \xi_j^{\mu*}$ is just a natural generalization of the Hebb-Hopfield factor $\xi_i^\mu \xi_j^\mu$ for (real) static patterns, which becomes the familiar outer-product form for complex vectors ξ.

2.2   Numerical Simulations

If the learning rule (9), with a proper shape of the learning window $A(t)$ is used, then the network is able to memorize the encoded patterns as stable attractors of the dynamics, and to replay them selectively. We consider P periodic patterns ($\mu = 1..., P$, $j = 1, ..., N$) characterized, as in eq. (4), by frequencies ω_μ and random phases ϕ_j^μ, where $\xi_j^\mu = e^{i\phi_j^\mu}$ is the vector of phase shift among units in pattern μ. Numerical simulations of eqs. (1,2) where J_{ij} is given by eq. (9) is shown in Fig. 1. Two periodic patterns $L^\mu(t)$ ($\mu = 1,2$) with random phases and frequencies $\omega_1 = 0.03ms^{-1}$, $\omega_2 = 0.09ms^{-1}$ have been memorized in a network of N=10000 units according to the learning rule (9), by using learning window shown in Fig.4. This gives the factor $\tilde{A}(\omega_\mu) = |a^\mu|e^{i\varphi_\mu}$, with $|a^1| = 1.899$,

$|a^2| = 1.760, \varphi_1 = -0.27\pi, \varphi_2 = -0.1\pi$ and $b = 0$. The factors $\tilde{A}(\omega_\mu), \mu = 1, 2$ are chosen so that both patterns become stable attractors of the dynamics (see section 2.3). When we carry out the numerical integration of the dynamics, under the initial condition $x_i(0) = L_i^\mu(0)$ with $\mu = 1$, the first pattern is retrieved, and, analogously the second is retrieved with initial condition close to the second pattern. Figure shows that the phase relationship among units in Fig. 1(A) are well reproduced in the collective oscillation during replay, shown in Fig.1(C); while the details of the wave forms and the frequency are different. In this sense, initial condition $x_i(0) = L_i^\mu(0)$ leads to retrieval (i.e. replay) of pattern L^μ, since the activity preserve the encoded phase relationship among units. The same sequence (i.e. the same phase shifts relationship among units) is replayed, but at a different oscillation frequency. Note indeed that the time-axis scale of Fig. 1(A) is different from that of Fig. 1(C). Later analysis will show that the relationship among the frequency ω_μ encoded into J_{ij} and the frequency of the network dynamics $\tilde{\omega}^\mu$ depends both from the single unit time-constant τ_d and, notably, from the shape of learning window used to learn J_{ij}. Figs.1(C) and 1(D) show also that the frequency of replay of pattern $\mu = 1$ is different of frequency of replay of pattern $\mu = 2$, since indeed the two encoding frequencies were different.

To understand analytically this behavior, and the relation among stability of attractors and frequencies, we define an order parameter which measure the overlap between the network activity $x_i(t)$ and the pattern $L^\mu(t)$, as the scalar product

$$m^\mu = \frac{1}{N} \sum_{i=1}^N \xi_i^\mu x_i \tag{11}$$

i.e. the overlap between the network activity and the encoded pattern of phase shifts. Order parameters m^μ are complex-valued which exhibits periodic oscillation on complex plain (see Figs.1(E) and 1(F)). The amplitude of oscillation of the order parameters is high when pattern is replayed, and is zero for not-replayed pattern. Replay frequency $\tilde{\omega}^\mu$ takes a different value from encoded pattern frequency ω_μ since replay occurs at a time scale different from encoded patterns.

2.3   Order Parameter Dynamics

Following [2] we rewrite local field (2) in eq. (1) in terms of order parameters m^μ and X,

$$h_i = \sum_\mu \text{Re}\left(\tilde{A}(\omega_\mu)\xi_i^\mu m^{\mu*}\right) + bX. \tag{12}$$

where X is the the mean activity

$$X = \frac{1}{N} \sum_i x_i. \tag{13}$$

and overlaps $m^\mu = (1/N)\sum_i \xi_i^\mu x_i$ as in eq.11.

A: encoded pattern $L^1(t)$ B: encoded pattern $L^2(t)$

C: replay of first pattern   D: replay of second pattern

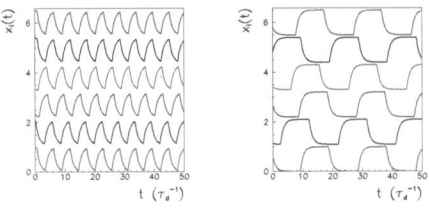

E: order parameters m^1 and m^2 F: order parameters m^1 and m^2

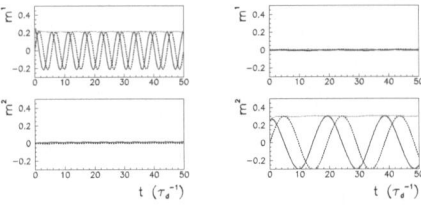

Fig. 1. In this numerical simulation, two periodic patterns $L^\mu(t)$ ($\mu = 1,2$) with random phases and frequencies $\omega_1 = 0.03ms^{-1}$, $\omega_2 = 0.09ms^{-1}$ have been memorized in a network of N=10000 units according to the learning rule (9), by using learning window in Fig. 4 which gives factors $\tilde{A}(\omega_\mu) = |a^\mu|e^{i\varphi_\mu}$, with $|a^1| = 1.760, |a^2| = 1.899$, $\varphi_1 = -0.27\pi$, $\varphi_2 = -0.1\pi$ and $b = 0$. The factors $\tilde{A}(\omega_\mu), \mu = 1, 2$ are such that both patterns are stable attractors of the dynamics (see section 3). (A,B): The first six units of the two encoded pattern $L^\mu(t)$ are plotted as a function of time. Activities of the six units of first pattern, $L_i^\mu(t)$, $i = 1, .., 6$, $\mu = 1$, are shown in A, while the second pattern, $L_i^\mu(t)$, $i = 1, .., 6$, $\mu = 2$, is shown in B. (C,D): The behavior of first six neurons x_i in the numerical simulation of eq. (1) are plotted as a function of time. The numerical integration of equation (1) are computed with the transfer function F(h) given by $F(h) = \frac{1}{2}(1+tanh(\beta h))$ with sufficient large β ($\beta = 100$). Initial condition set to $x_i(0) = L_i^\mu(0)$ with $\mu = 1$ induces the retrieval (replay) of pattern $\mu = 1$, as shown in (C), while initial condition $x_i(0) = L_i^\mu(0)$ with $\mu = 2$. induces the retrieval (replay) of second pattern, as shown in (D). (E,F): The time evolutions of order parameters m^μ, with $\mu = 1, 2$, defined by eq. (11) are plotted as a function of time. The real (red line) and the imaginary part (dashed blue line) of m^μ oscillate in time, in agreement with predictions. Module (dotted pink line) is constant in time. Fig. (E) refers to initial condition as in (C) with replay of pattern $\mu = 1$, while fig. (F) correspond to picture (D) with replay of pattern $\mu = 2$. Colors are in the online version of the paper.

By representing the overlap m^μ in the polar form

$$m^\mu = |m^\mu|e^{i\theta^\mu}. \tag{14}$$

and taking the limit $N \to \infty$ for finite P to replace the summation over i by the average $\langle . \rangle$ defined by $\langle f(\xi^1, .., \xi^P) \rangle = (\frac{1}{2\pi})^P \int_0^{2\pi} d\phi^1 .. \int_0^{2\pi} d\phi^P f(e^{i\phi^1}, .., e^{i\phi^P})$, as in [2], from eqs. (1,2, 13,14) we obtain a set of nonlinear differential equations for the module and the phase of overlap m^μ, and mean activity X:

$$\tau_d \frac{d}{dt}|m^\mu| = -|m^\mu| + \left\langle \text{Re}(\xi^\mu) F\left(\sum_\nu \text{Re}\left(\tilde{A}(\omega_\nu)\xi^\nu|m^\nu|\right) + bX\right)\right\rangle \tag{15}$$

and

$$\tau_d \frac{d}{dt}\theta^\mu = \frac{1}{|m^\mu|}\left\langle \text{Im}(\xi^\mu) F\left(\sum_\nu \text{Re}\left(\tilde{A}(\omega_\nu)\xi^\nu|m^\nu|\right) + bX\right)\right\rangle. \tag{16}$$

$$\tau_d \frac{d}{dt}X = -X + \left\langle F\left(\sum_\nu \text{Re}\left(\tilde{A}(\omega_\nu)\xi^\nu|m^\nu|\right) + bX\right)\right\rangle. \tag{17}$$

Equation (15) describes the dynamics of absolute value of overlap m^μ, while $d\theta^\mu/dt$ in eq. (16) gives the frequency of oscillation of overlap m^μ. Equations (15) and (17) yield $(P+1)$-dimensional closed dynamics for the order parameters $|m^\mu|$ $(\mu = 1, \ldots, P)$ and X.

To investigate the properties of this order parameters dynamics, we analyze solutions of dynamics (15) and (17) corresponding to successful replay of one encoded pattern. Safely we can assume pattern with $\mu = 1$ to be the retrieved one. In the successful pattern retrieval (replay), as the one described in Fig. 1, overlap $|m^1|$ eventually settles into the stationary state with the constant absolute value, while other overlaps $m^\mu(\mu = 2, \ldots, P)$ remain close to zero. We mathematically define this retrieval state as

$$|m^1| \neq 0, \tag{18}$$
$$|m^2| = |m^3| = \ldots = |m^P| = 0, \tag{19}$$

and

$$\frac{d\theta^1}{dt} \neq 0 \tag{20}$$

$$\frac{d}{dt}|m^1| = \frac{d}{dt}X = 0, \tag{21}$$

The retrieval frequency $\tilde{\omega}^1$, namely, the oscillation frequency of overlap in retrieval state $\mu = 1$, is

$$\tilde{\omega}^1 = \frac{d\theta^1}{dt}, \tag{22}$$

Solutions of eqs. (15-17), corresponding to retrieval of pattern $\mu = 1$, are:

$$X = \left\langle F\left(\text{Re}\left(\tilde{A}(\omega_1)\xi^1 |m^1| \right) + bX \right) \right\rangle$$
$$= \frac{1}{2\pi} \int_0^{2\pi} F\left(\left|\tilde{A}(\omega_1)\right| |m^1| \cos\left(\phi^1\right) + bX \right) d\phi^1. \tag{23}$$

$$|m^1| = \left\langle \text{Re}\left(\xi^\mu\right) F\left(\sum_\nu \text{Re}\left(\tilde{A}(\omega_\nu)\xi^\nu |m^\nu| \right) + bX \right) \right\rangle$$
$$= \cos\varphi_1 \frac{1}{2\pi} \int_0^{2\pi} \cos\phi^1 F\left(\left|\tilde{A}(\omega_1)\right| |m^1| \cos\phi^1 + bX \right) d\phi^1. \tag{24}$$

$$\tilde{\omega}^1 \equiv \frac{d\theta^1}{dt} = -\frac{\tan\varphi_1}{\tau_d} = -\frac{1}{\tau_d}\frac{\text{Im}(\tilde{A}(\omega_1))}{\text{Re}(\tilde{A}(\omega_1))}. \tag{25}$$

A solution of Eqs. (23,24,25) determines $|m^1|$ and X in retrieval state. These equations can be solved numerically for arbitrary form of function $F(h)$, and analytically in some cases. This analysis shows when there exists a solution of Eqs. (23,24,25) then it is a collective oscillation with frequency $\tilde{\omega}^1 = -\frac{\tan\varphi_1}{\tau_d}$, and overlap $|m^1|$ between $x(t)$ and pattern $L^1(t)$. Retrieval state, i.e. replay of encoded pattern, is a collective oscillation of the network's units, with a phase-shift relation among units equal to the phase-shifts relation of the encoded pattern; The collective oscillation during replay has a frequency $\tilde{\omega}^1$ which is related but not equal to the encoded pattern's frequency, and depends from the learning window shape (see eq. 25).

In some cases explicit analytical solution exists. When the nonlinear transfer function $F(h)$ is the Heaviside function of the form

$$F(h) = H(h) = \begin{cases} 0, h < 0 \\ 1, 0 \le h \end{cases}, \tag{26}$$

from Eqs. (23,24,25) one get

$$|m^1| = \frac{1}{\pi}\cos\varphi_1 \sin(\pi X),$$
$$sin(2\pi X) = -\frac{b}{\text{Re}(\tilde{A}(\omega_1))}2\pi X$$
$$\frac{d}{dt}\theta^\mu = \tilde{\omega}^1 = -\frac{\tan\varphi_1}{\tau_d} \tag{27}$$

where φ_1 is defined in eq. (8). The solution, when pattern 1 is retrieved, depends only from parameters of pattern 1, and in particular from $\tilde{A}(\omega_1)$, and do not depends on all others $\tilde{A}(\omega_\nu), \nu = 2, .., P$ (however the stability of this solution will depend from all $\tilde{A}(\omega_\nu), \nu = 1, .., P$, as it will be shown in the next section).

With $b = 0$ (balance between potentiation and depression in learning window $\int_{-\infty}^{\infty} A(t)dt = 0$), solutions simplify furthers and one obtains

$$
\begin{aligned}
X &= 1/2 \\
|m^1| &= (1/\pi)\cos\varphi_1 \\
\frac{d\theta^1}{dt} &= \tilde{\omega}^1 = -\frac{\tan\varphi_1}{\tau_d}
\end{aligned}
\tag{28}
$$

Eqs. (28) means that

$$
|m^1| = (1/\pi)\sqrt{\frac{1}{1 + (tan\varphi_1)^2}} = (1/\pi)\sqrt{\frac{1}{1 + (\tilde{\omega}^1\tau_d)^2}}
\tag{29}
$$

It means that the time decay constant of single unit dynamics τ_d determines the way at which order parameter decreases with increase of replay frequency $\tilde{\omega}^1$. When $|m|$ takes a low value, then $x_i(t)$ shows small amplitude oscillations around a flat activity 0.5, while largee m gives larger amplitudes oscillations aroung 0.5 (see also Fig.1(E) and 1(F)).

3   Multiple Frequencies and Stability

The solutions (23-25) yield fixed points of the dynamics (15-17). However, some of these solutions are unstable fixed points, which are useless for the purpose of stable associative memories. When a pattern is encoded as an unstable solution, during retrieval the network shows only a transient activation of that dynamical pattern (see Fig.2). On the contrary when a solution is a stable attractor of the dynamics, this solution is a spontaneous pattern of activity. This means that a pattern encoded as a stable solution can emerge and stay active also in total absence of external input, when the state of the network falls in the basin of attraction of that pattern.

In [2], we analyze linear stability of solutions of Eqs. (23-25), in the case of multiple patterns ($1 < P$) but with identical frequency $\omega_\mu = \omega_0$ for all $\mu = 1, .., P$. Notably, in the case of the Heaviside function, solution state was always stable for arbitrary finite value of P if b=0 and all patterns have same frequency. We show here that, when patterns have multiple frequencies, stability is not assured even if b=0, but condition of stability depends from the encoded frequencies.

When frequencies of encoded patterns are different, then the factor $\tilde{A}(\omega_\mu)$ which enters in the learning rule $J_{ij} = \frac{1}{N}\sum_\mu Re[\tilde{A}(\omega_\mu)\xi^\mu\xi^{\mu*}] + \frac{b}{N}$ is not the same for all patterns. This factor $\tilde{A}(\omega_\mu)$ have influence both on the size of order parameter during retrieval solution $|m^1|$, and also on the stability of such solutions. Numerical simulations show that when we encode multiple patterns with different frequencies each pattern can have a different stability. Fig.2 shows numerical simulation results when we encode two oscillatory patterns with random phases and two different frequencies, with factors $\tilde{A}(\omega_\mu) = |a^\mu|e^{i\varphi_\mu}$ such that pattern $L^2(t)$ is stable, and the pattern $L^1(t)$ is unstable. Fig. 2 shows that even if we try to recall pattern 1, after a short transient the network activity goes far from pattern 1, and pattern 2 (the stable one) is replayed. If pattern 2 is not encoded into the network, and frequency of pattern 1 is the only frequency encoded into the net, then pattern 1 is a stable attractor. It is pattern 2 which make pattern 1 to become unstable.

To understand this behavior and find the analytical condition for stability, we extend here the linear stability analysis to the case when multiple patterns with multiple frequencies are encoded.

To study stability, we have to study dynamics of deviations of order parameters from the fixed point in eq. (18-21), i.e. perturbations of retrieved pattern $\delta|m^1|$ as well as perturbations of non-retrieved patterns $\delta|m^\mu|(\mu = 2,\ldots,P)$ and δX. From eqs. (15-17, 23-25), dynamics of deviations $\delta|m^\mu|$ and δX are given by

$$
\tau_d \frac{d}{dt}
\begin{pmatrix}
\delta X \\
\delta|m^1| \\
\delta|m^2| \\
\vdots \\
\delta|m^P|
\end{pmatrix}
= \mathbf{B}
\begin{pmatrix}
\delta X \\
\delta|m^1| \\
\delta|m^2| \\
\vdots \\
\delta|m^P|
\end{pmatrix}
\tag{30}
$$

with

$$
\mathbf{B} =
\begin{pmatrix}
-1 + I(1,b) & I(1,a^1\xi^1) & I(1,a^2\xi^2) & \cdots & I(1,a^P\xi^P) \\
I(\xi^1,b) & -1 + I(\xi^1,a^1\xi^1) & I(\xi^1,a^2\xi^2) & \cdots & I(\xi^1,a^P\xi^P) \\
I(\xi^2,b) & I(\xi^2,a^1\xi^1) & -1 + I(\xi^2,a^2\xi^2) & \cdots & I(\xi^2,a^P\xi^P) \\
\vdots & \vdots & \vdots & \ddots & \vdots \\
I(\xi^P,b) & I(\xi^P,a^1\xi^1) & I(\xi^P,a^2\xi^2) & \cdots & -1 + I(\xi^P,a^P\xi^P)
\end{pmatrix}
\tag{31}
$$

where we use an abbreviation

$$
I(f,g) = \left\langle \operatorname{Re}(f) F' \left(\operatorname{Re}\left(a^1\xi^1|m^1|\right) + bX \right) \operatorname{Re}(g) \right\rangle.
\tag{32}
$$

From definition of $I(f,g)$ in eq. (32), it is straightforward to show

$$
I\left(1, a\xi^\mu\right) = I\left(\xi^1, a\xi^\mu\right) = I\left(\xi^\mu, b\right) = I\left(\xi^\mu, a\xi^1\right) = 0,
$$
$$
2 \le \mu,
\tag{33}
$$

and

$$
I\left(\xi^\mu, a^\nu\xi^\nu\right) = 0, \qquad 2 \le \mu,\nu,\ \mu \ne \nu.
\tag{34}
$$

We can rewrite matrix $\mathbf{B}$ in the form

$$
\mathbf{B} =
\begin{pmatrix}
\mathbf{A} & & & \mathbf{0} \\
& -1 + I(\xi^2,a^2\xi^2) & & \\
& & \ddots & \\
\mathbf{0} & & & -1 + I(\xi^P,a^P\xi^P)
\end{pmatrix},
\tag{35}
$$

where $\mathbf{A}$ represents the matrix defined by

$$
\mathbf{A} =
\begin{pmatrix}
-1 + I(1,b) & I\left(1,a^1\xi^1\right) \\
I\left(\xi^1,b\right) & -1 + I\left(\xi^1,a^1\xi^1\right)
\end{pmatrix}.
\tag{36}
$$

A: units' activity   B: order parameters m^1 and m^2

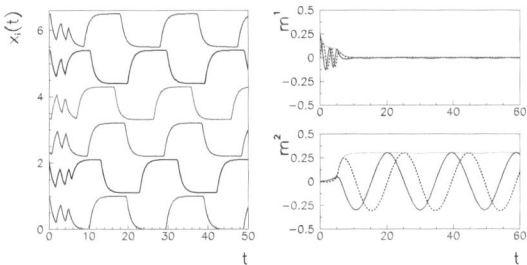

C overlap $|m^1|$ versus $-\varphi_1/\pi$

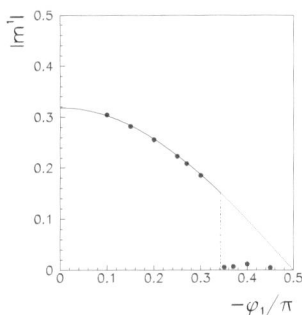

Fig. 2. (A,B) In this numerical simulation, two periodic patterns $L^\mu(t)$ ($\mu = 1,2$) with random phases and different frequencies have been encoded in a network of N=10000 units. The factor $\tilde{A}(\omega_\mu)$ of the two patterns are such that only pattern 2 is a stable attractor, while pattern 1 is unstable. In particular it's $\tilde{A}(\omega_\mu) = |a^\mu|e^{i\varphi_\mu}$ with $|a^1| = 1, \varphi_1 = -0.4\pi$ for pattern 1, and $|a^2| = 1, \varphi_2 = -0.1\pi$ for pattern 2, such that $Re(\tilde{A}(\omega_1)) < Re(\tilde{A}(\omega_2))/2$ and stability condition is not satisfied for pattern 1. The activities of the first six units is shown, when initial condition is $x_i(0) = L^1(0)$. Although the initial condition has high overlap with pattern 1 and very small overlap with patter 2, the network activity, after a short transient, replay patter 2, because pattern 1 is unstable and 2 is stable. In (B) the time evolutions of the order parameters m_1 and m_2 during the same simulation is shown. The real (red line), the imaginary part (dashed blue line), and the module (dotted pink line) of m^1 (upper inset) and m^2 (lower inset) are shown as a function of time. The overlap $|m^1|$ of the unstable pattern 1 decays after a short transient. Colors are in the online version of the paper. In (C) the overlap $|m_1|$ (after the short transient) is shown as a function of parameter $-\varphi_1/\pi$, while pattern 2 is encoded with $\varphi_2 = -0.1\pi$ as in (A,B). We see, in agreement with analytical calculations, that a critical value appears ($-\varphi_1/\pi = 0.34$, corresponding to $Re(\tilde{A}(\omega_1)) = Re(\tilde{A}(\omega_2))/2$), such that for lower values the pattern 1 is a stable attractor, while for higher values the pattern 2 make pattern 1 unstable.

We denote eigenvalues of matrix $\mathbf{A}$ by λ_0 and λ_1. Therefore, matrix $\mathbf{B}$ has the same eigenvalues as matrix $\mathbf{A}$, plus $P - 1$ eigenvalues

$$\kappa^\nu = -1 + I\left(\xi^\nu, a^\nu \xi^\nu\right) \quad \nu = 2, .., P \tag{37}$$

The solution in Eqs. (23-25), corresponding to the retrieval of pattern 1, is stable when the eigenvalues satisfy all the stability conditions:

$$\mathrm{Re}\left(\lambda_l\right) < 0 \quad \text{and} \quad \mathrm{Re}\left(\kappa^\nu\right) < 0, \qquad l = 0, 1. \qquad \nu = 2, .., P \tag{38}$$

While stability of solutions with a single encoded pattern ($P = 1$) is determined only from λ_0 and λ_1, stability with multiple encoded patterns ($1 < P$) requires further evaluation of all $P - 1$ eigenvalues κ^ν. Note that eigenvalue κ^ν contains both $a^\nu = \tilde{A}(\omega_\nu)$ and $a^1 = \tilde{A}(\omega_1)$ in its definition, and indeed stability of pattern 1 depends from all encoded frequencies ω_ν.

Let us apply the above analysis to the case when transfer function $F(h)$ is the Heaviside function eq.(26), in such a case we can evaluate analitically the condition of stability.

Using the Heaviside function as a transfer function $F(h) = H(h)$, and consequently $F'(h) = \delta(h)$, it's:

$$I(\xi^1, a^1\xi^1) = c_1 2|a^1|cos^2(\pi X)cos(\varphi_1)$$

$$I(1, a^1\xi^1) = c_1 2|a^1|cos(\pi X)$$

$$I(1, b) = c_1 2b, \qquad I(\xi^1, b) = c_1 2bcos(\pi X)cos(\varphi_1)$$

$$I(\xi^\nu, a^\nu\xi^\nu) = c_1|a^\nu|cos(\varphi_\nu)$$

where

$$c_1 = 1/(2|a_1|cos(\varphi_1)sin^2(\pi X)). \tag{39}$$

then, eigenvalues λ_0, λ_1 and κ^ν of matrix B are

$$\lambda_0 = -1$$

$$\lambda_1 = -1 + \frac{|a^1|cos^2(\pi X)cos(\varphi_1) + b}{|a^1|sin^2(\pi X)cos(\varphi_1)}$$

$$\kappa^\nu = -1 + \frac{|a^\nu|cos(\varphi_\nu)}{2|a^1|cos(\varphi_1)sin^2(\pi X)} \tag{40}$$

In the case when encoded patterns have all the same frequency, i.e. $\omega_\mu = \omega_0$ discussed in [2], all eigenvalues κ^ν are degenerate, and for example, when $b = 0$ we obtain $\kappa^\nu = -\frac{1}{2}$, and retrieval solution is always stable, independently from number of encoded patterns P.

On the contrary in the case when encoded patterns have distributed frequencies, i.e. ω_μ may depend from index μ, then stability condition (38) is not always satisfied, even if b=0. Indeed, from (40), stability condition $\kappa^\nu < 0$ requires:

$$|a^1|cos(\varphi_1) > \frac{|a^\nu|cos(\varphi_\nu)}{2sin^2(\pi X)} \qquad \nu = 1, .., P \qquad (41)$$

In particular when $b = 0$, i.e. when there is a balance between depression and potentiation such that $\int_{-\infty}^{\infty} A(t)dt = 0$, then $\lambda_{0,1} = -1$ and stability conditions (38,40) become simply:

$$|a^1|cos(\varphi_1) > \frac{|a^\nu|cos(\varphi_\nu)}{2}, \quad \nu = 1, .., P \qquad (42)$$

where a^ν, φ_ν, defined in (8), depends from Fourier transform of learning window at frequency ω_ν.

This means that when we encode patterns with P different frequencies, $\omega_1, \omega_2, .., \omega_P$, then these patterns are stable attractors only if the Fourier transform of the learning window evaluated at these P frequencies satisfy the condition

$$Re\tilde{A}(\omega_\mu) > \frac{Re\tilde{A}(\omega_\nu)}{2} \qquad \forall \nu = 1, .., P \text{ and } \forall \mu = 1, .., P \qquad (43)$$

It means that the P frequencies can be encoded as stable attractors in the same network only if the following equation is satisfied:

$$Re\tilde{A}(\omega_\mu) \in [\frac{c_{max}}{2}, c_{max}] \quad \forall \mu = 1, .., P \qquad (44)$$

where c_{max} is the $\max_\mu Re(\tilde{A}(\omega_\mu))$, i.e., each value $Re(\tilde{A}(\omega_\mu))$ is larger then half of the maximum value c_{max}.

Each time that one of the values, let us say $Re(\tilde{A}(\omega_\nu))$ is lower then half of the maximum value c_{max}, the pattern encoded at that frequency ω_ν is unstable.

Numerical simulations confirm this behavior. In Fig. 2(C) results are shown when two patterns with two different frequencies have been encoded with factors $\tilde{A}(\omega_1) = e^{i\varphi_1}$ and $\tilde{A}(\omega_2) = e^{i\varphi_2}$ (Note that here we artificially put $\tilde{A}(\omega_1) = e^{i\varphi_1}$ and $\tilde{A}(\omega_2) = e^{i\varphi_2}$ to check the validity of the present stability analysis, later we use $A(\tau)$ explicitly). The order parameter $|m_1|$ is shown as a function of $-\varphi_1/\pi$, when φ_2 is fixed ($\varphi_2 = -0.1\pi$). Numerical simulation results (dots) are in good agreement with prediction (line). They shows that at the point $\varphi_1 = -0.34\pi$ such that $cos(\varphi_1) = cos(\varphi_2)/2$ there is a transition, from a retrieval state with $m_1 > 0$ to a state with zero overlap $m_1 = 0$. The pattern 2 make pattern 1 unstable when $Re\tilde{A}(\omega_1) < \frac{Re\tilde{A}(\omega_2)}{2}$. The behavior of the network when $\varphi_1 = -0.4\pi, \varphi_2 = -0.1\pi$, such that pattern 1 is unstable, is shown in Fig. 2.

These results imply that the possibility of the network to work at different frequencies, and the set of frequencies encodable as attractors in a same network, depends critically from the shape of learning window though eq. (44). In the following paragraph we will examine the implications of particular shapes of learning window on network dynamics.

3.1   Learning Window Shape

From eqs. (23-25) and from stability conditions of retrieval solutions, (38,40), one can see that shapes of learning windows where potentiation and depression balance each other such that $\int_{-\infty}^{\infty} A(t)dt = 0$ and therefore $b = 0$, maximize capacity. Indeed, using the Heaviside function, for $b \equiv 2P \int_{-\infty}^{\infty} A(t)dt \neq 0$, there is a critical number of patterns P_c (see also [2]), after which self-consistent solutions (23-25) do not exist or are unstable with respect to λ_1 or k^ν; while for $b = 0$ retrieval solutions exist, λ_0 and λ_1 satisfy the conditions (38), for each finite value of P, and the only remaining proble is the stability condition (44).

Let us analyze further how learning window shape, where potentiation and depression balance each other such that $b = 0$, determines the dynamics of the attractor network. First of all, as shown in eq. (25), when a pattern is retrieved, the frequency at which it is replayed depends explicitly from learning window $A(t)$ shape, and in particular is equal to $\tilde{\omega}^1 = -\frac{tan(\varphi_1)}{\tau_d}$, where φ_1 is the argument of learning window's Fourier transform evaluated at the encoding frequency. Also the size of order parameter $|m^1|$ in retrieval state, which measures the overlap between dynamics and the phase-relationships of the encoded patterns, depends on the Fourier transform of learning window evaluated at the encoding frequency (see eq.24). In the case in which the Heaviside function is chosen as F in (1), and $b = 0$, then the size of $|m_1|$ do not depend from the module of $\tilde{A}(\omega_1)$ but only from its argument (see eq.(28)).

Asymmetry of the shape of learning window $A(\tau)$ with respect to $\tau \to -\tau$ influence both retrieval state and its stability. Clearly $\text{Im}\tilde{A}(\omega_\mu) = 0$ if $A(\tau)$ is symmetric in τ and $\text{Re}\tilde{A}(\omega_\mu) = 0$ if $A(\tau)$ is antisymmetric. Both these extreme cases fails to encode and replay properly oscillatory patterns. When $A(\tau)$ is symmetric, then replay frequency $\tilde{\omega}^1$ is zero (static output, no oscillation), while in the limit of $A(\tau)$ perfectly antisymmetric the replay frequency tends to infinity but with overlap $|m^1|$ which tends to zero.

Notably, Abarbanel et al in a recent paper [51] introduced and motivated an function of $A(\tau)$, which gives a very good fit of different results. To understand the implication of experimentally plausible learning window shapes of STDP, we study the following function, taken from the one introduced and motivated by Abarbanel et al [51],

$$A(\tau) = \begin{cases} a_p e^{-\tau/T_p} - a_D e^{-\eta\tau/T_p} \ for \ \tau > 0 \\ a_p e^{-\tau/T_D} - a_D e^{-\eta\tau/T_D} \ for \ \tau < 0 \end{cases} \tag{45}$$

We use the same parameters taken in [51] to fit the data of Bi and Poo [30],

$$a_p = \frac{\gamma}{1/T_p + \eta/T_D} \tag{46}$$

$$a_D = \frac{\gamma}{1/T_D + \eta/T_D} \tag{47}$$

with $T_p = 10.2ms, T_D = 28.6ms, \eta = 4, a_p = 177, a_D = 98$. Notably, the function (45) automatically satisfies the condition

A: Learning window $A(\tau)$ B: $|m^1| = cos(\varphi_1)/\pi$ vs. ω_1

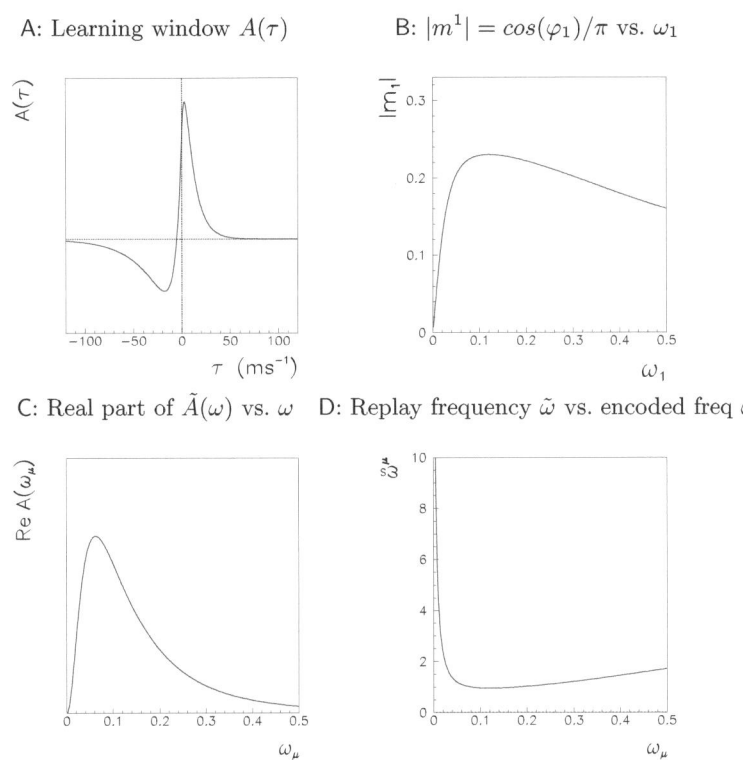

C: Real part of $\tilde{A}(\omega)$ vs. ω D: Replay frequency $\tilde{\omega}$ vs. encoded freq ω

Fig. 3. The learning window in eq. (45), with parameters which fit the experimental data of Bi and Poo [30], is shown in (A) In (B) it's shown retrieval's overlap $|m_1|$ as a function of encoding frequency $|\omega_1|$ when $A(t)$ in (A) is used, and transfer function is the Heaviside function $F(h) = H(h)$. In (C) the real part of the Fourier transform of learning window, i.e. $Re\tilde{A}(\omega)$, is shown as a function of encoded frequency ω. This is relevant to establish the set of frequencies that can be encoded as stable attractors in the same network (see stability condition (38,44)). We plot in (D) the replay frequency $\tilde{\omega}$ (see eq. (22,25)) as a function of the encoded frequency ω when the learning window shown in (A) is used. The frequency of replay in (D) is measured in units of τ_d^{-1}, while the encoded frequency is measured in ms^{-1}.

$$\int_{-\infty}^{\infty} A(\tau)d\tau = 0 \tag{48}$$

i.e. $\tilde{A}(0) = 0$ and $b = 0$.

From eq. (25) we get Fig. 3(D), showing the frequency of replay measured in units of τ_d, as a function of the encoded frequency measured in ms^{-1}, when using the learning window (45) in Fig. 3(A). Numerical simulations confirm that the time-scale of the retrieval depends from the time-constant of single unit dynamics τ_d, and from the connectivity via the Fourier transform of $A(t)$, in agreement with Fig. 3(D). When a pattern with frequency ω is encoded into the

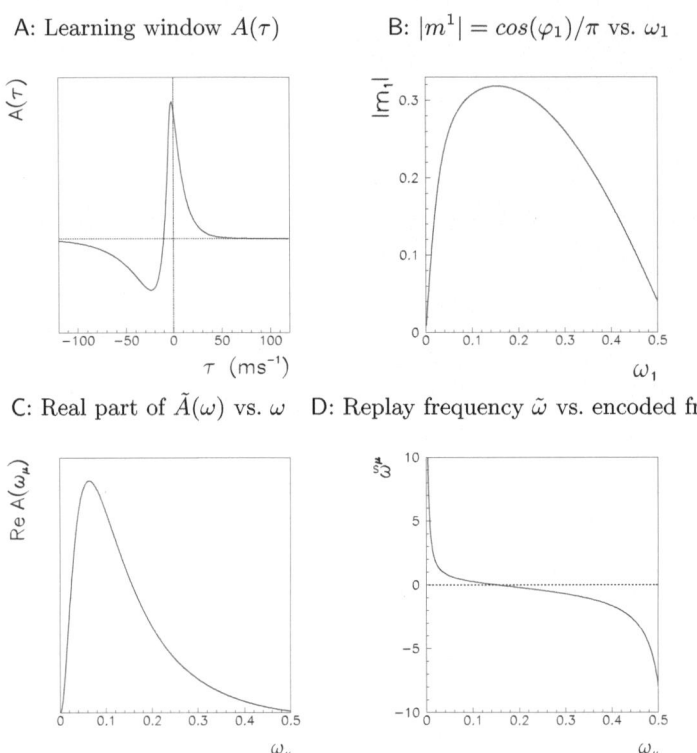

A: Learning window $A(\tau)$

B: $|m^1| = cos(\varphi_1)/\pi$ vs. ω_1

C: Real part of $\tilde{A}(\omega)$ vs. ω

D: Replay frequency $\tilde{\omega}$ vs. encoded freq ω

Fig. 4. The learning window in eq.(45) but time shifted of 5 ms, is shown in (A) Same parameters of Fig 3(A), except the time shift of 5 ms. The overlap $|m^1|$ and the real part of Fourier transform of learning window are shown in B and C. We plot in (D) the replay frequency $\tilde{\omega}$ as a function of the encoded frequency ω when the learning window shown in (A) is used. The frequency of replay in (D) is measured in units of τ_d^{-1}, while the encoded frequency is measured in ms^{-1}.

network, using the learning window shown in Fig.3(A) and the Heaviside transfer function, the order parameter size $|m| = \frac{cos(\varphi)}{\pi}$ of the retrieval solution is shown in Fig.3(B) as a function of encoded frequency ω. When multiple frequencies are encoded into the same network condition (44) should be evaluated to check stability of encoded patterns. From Fig.3(C) and eq. (44) it is possible to evaluate which intervals of frequencies ω can be encoded as stable attractors in the same network.

However a change in the shape of the learning window $A(\tau)$ changes significantly the Fourier transform $\tilde{A}(\omega)$, and therefore change significantly the dynamics. Here we show the case in which the learning window in eq. (45) is time-shifted on the left of $\tau_0 = 5ms$, i.e. $\tau \to \tau + \tau_0$. Figure 4 shows the learning window, and the quantities related to its Fourier transform, i.e. the overlap $|m|$ of retrieved pattern, the $Re\tilde{A}(\omega)$, and replay frequency $\tilde{\omega}$ versus encoded frequency ω. We note that, with respect to Figk. 3, in Fig 4 a larger interval of frequency

A: Encoded pattern

C: Encoded pattern with sorted units

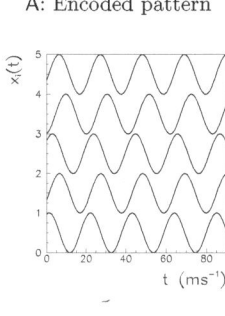

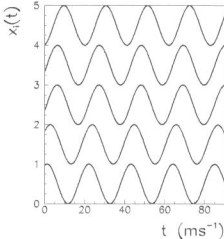

B: Replay

D: Replay with sorted units

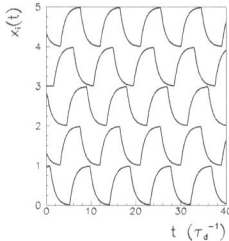

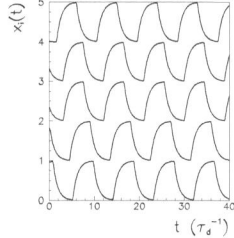

E: Order parameter m^1 versus time

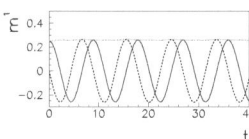

Fig. 5. In this numerical simulation, a periodic pattern $L^1(t)$ with random phases and frequency $\omega^1 = 0.4$ has been memorized in a network of N=10000 units according to the learning rule (9) by using the time window $A(\tau)$ shown in Fig. 4 (eq.(45) but with a time shift of $5ms$). The first six units of pattern $L^1(t)$ are plotted in (A) as a function of time. In (B) The evolution of first six units $x_i(t)$ in the numerical simulations of eq. (1) are plotted as a function of time. The numerical integration of equation (1) is computed with the transfer function $F(h) = \frac{1}{2}(1 + \tanh(\beta h))$ with $\beta = 100$. Initial condition $x_i(0) = L_i^1(0)$ induces the retrieval (replay) of pattern $\mu = 1$. However since $\omega^1 = 0.4$ gives $\varphi_1 = 1.02$, it's $\tilde{\omega}^1 = -\frac{1.6}{\tau_d}$ negative, and the pattern is replayed in a reversed order. To see the correspondence between the encoded pattern and the replayed one, we sorted the units with respect to the phase of encoded pattern ϕ_i^1. In (C) the encoded pattern shown in (A) is shown with sorted order. In (D) the activity of the network during replay shown in (B) is shown using the same sort of units used in (C). Note that the sequence during replay is reversed. (E) The time evolution of order parameter m^1, is plotted as a function of time. The real (red line) and the imaginary part (dashed blue line) of m^1 oscillate in time, in agreement with predictions. Module (dotted pink line) is constant in time.

ω gives low $\tilde{\omega}$ and therefore large overlap $|m|$. Moreover in Fig.4(D) we see that there is a region where $\tilde{\omega}$ takes a negative value, and therefore negative replay frequencies are possible. In the case of negative replay frequencies, the network replays encoded patterns in the reversed order with respect to time. This is shown in Fig. 5, showing numerical simulations of the network of eq. (1), when learning, with the learning window in Fig.4, a periodic pattern with frequency $\omega^1 = 0.4$. Since $\omega^1 = 0.4$ gives $\varphi_1 = 1.02$, and therefore $\tilde{\omega}^1 = -\frac{1.6}{\tau_{d}}$ is negative, this means that the encoded pattern is replayed in a reverse order in time. This means that the relative phases during replay are opposite with respect to phases of encoded pattern used during learning. To see the correspondence between the encoded pattern and the replayed one, we sorted the units with respect to the phase of encoded pattern ϕ_i^μ, so that sorted units shows increasing phases in the encoded pattern. Using this sorted units the replay shows decreasing phases, i.e. the sequence is reversed. The frequency and the order at which the phases ϕ_i of the encoded pattern are recalled is related to φ. Eq.(25) shows that, when $b = 0$ and $F(h) = H(h)$, forward replay occurs when $-\frac{\pi}{2} < \varphi < 0$ and replay in reversed order when $\frac{\pi}{2} > \varphi > 0$. Therefore, in order to get retrieval of a phase-coded memory in reversed order with respect to the order experienced during learning, the learning window and the frequency during learning yield the Fourier Transform with $\frac{\pi}{2} > \varphi > 0$.

4   Conclusions

We analyzed the possibilities to encode in the network multiple spatio-temporal oscillatory patterns, with different frequencies, in such a manner that the network is able to selectively recall one of them, even in total absence of any external input, depending from the initial state of the network. Each encoded pattern is a collective oscillation, the frequency of the oscillation is the same for all units of the network for a given pattern (but may be different from pattern to pattern), while the phase of the oscillation are randomly chosen in $[0, 2\pi]$, and therefore are different for each unit of the network and from pattern to pattern.

We found that

1) Using a STDP-based learning rule (3), the patterns encoded are stable dynamical attractors only if their frequencies satisfy some conditions which depend on the shape of the learning window. Therefore, the shape of the learning window influences the possibility of coexistence of different frequencies as multiple stable attractors in the same network. When a pattern is encoded as an unstable solution, during retrieval the network shows only a transient activation of that dynamical pattern (see Fig.2). On the contrary when a solution is a stable attractor of the dynamics, this solution is a spontaneous pattern of activity. This means that a pattern encoded as a stable solution can emerge and stay active also in total absence of external input, when the state of the network falls in the basin of attraction of that pattern. Numerical simulations confirm analytical predictions.

2) During recall, the phases of firing of all units relative to the ongoing rhythm match the phases of the encoded pattern, while the frequency of the collective

oscillation is different from the one of the encoded pattern. The new frequency depend from the shape of the learning window, the frequency of encoded pattern, and the time constant of the single unit dynamics. It means that the same pattern can be replayed at a different time-scale, compressed in time or dilated in time, depending on the value of the Fourier transform of the learning window at encoding frequency, $\tilde{A}(\omega_\mu)$, and on τ_d.

The computational role and functional implications of the STDP plasticity, have been explored recently from many points of view (see for example [40,38,39, 51,48,41,20,18,25] and reference therein). In particular in the paper [48] authors discuss how the gap between synaptic and behavioral timescales can be bridged with STDP if stimuli to be associated generate sustained responses that decay slowly appropriately in time. The rule (3) we use in this paper share similarity with the analitical model of STDP used in [48]. While in our work, rule (3) is used to make the network to work as associative memory for dynamical attractors, in the work [48] authors discuss how it can generate connections strength that vary with time-delay between behavioral signals on the time scale of seconds, even though STDP is sensitive to correlations between pre- and post-synaptic activity on time-scale of tens of milliseconds.

Differently from the work [25], where this rule is used to make a network of coupled excitatory and inhibitory units become resonant to the encoded oscillatory patterns, and relax to the encoded pattern when a corrupted version of the pattern is presented as input, here (and in [2]) we study conditions under which the network can encode these patterns as spontaneous patterns of activity, i.e. patterns which emerges in total absence of external inputs, when the current state of the network, for any reason, is such to fall in the basin of attraction of encoded pattern. For example, it may be that an external input forces the network state to be such that one attractor is recalled, then the activity corresponding to that attractor is spontaneously active and self-sustained, in absence of external input. In the past years, increasing experimental evidence has been accumulated as to the significance of spontaneous ongoing network activity in different brain regions [52,53,54]. Our model support spontaneous activities to be the encoded pattern of collective oscillations, and it allows the existence of many spontaneous oscillatory patterns in the same network, even with different frequencies.

In our model the learning rule (3), after encoding an oscillatory pattern with frequency ω_μ and phase ϕ_j^μ on unit j, bring to a connectivity shown in eq. (7) where $|a^\mu|e^{i\varphi_\mu}$ is the Fourier transform of learning window $A(\tau)$ at frequency ω_μ, and $\tilde{A}(0)$ is zero when there is a balance between potentiation and depression in the learning window shape. When multiple phases-coded memories, each memory with a different frequency ω^μ and different phases ϕ_i^μ on each unit i, are encoded, then synaptic connections in our framework are given by eq. (9) The connections in eqs. (7) and (9), written in terms of phase of firing with respect to the ongoing rhythm, share same similarity with learning rule discussed in a Q-state clock neural network model [42], with learning rules used in the phase-oscillators frameworks [43, 13, 44], and with the STDP-based rule used in [41] and in spiking models [19, 18].

In our framework the frequency at which the phases ϕ_i^μ are recalled is related to φ^μ. Eq.(25) predicts forward replay when $-\frac{\pi}{2} < \varphi^\mu < 0$ and replay in reversed order when $\frac{\pi}{2} > \varphi^\mu > 0$. The size of overlap $|m^\mu|$ by which phases are replayed also is related to φ^μ (see eq. (28)). Moreover in order to store properly the phase-coded memory, as stable attractor, $|a^\mu|$, φ^μ and ω_μ should satisfy stability conditions (38,40). Under the conditions studied here, different phase-coded memories, each with a different frequency, can be recalled selectively, at different frequencies, in forward ($-\pi < \varphi^\mu < 0$) or reverse order ($\pi > \varphi^\mu > 0$).

Recently, forward replay and reverse replay of spatiotemporal neural activities has been reported in the various regions in the brain [9, 10, 11, 12]. The possible relation of the forward replay to the STDP has been pointed out in the previous study regarding the sharp wave burst in the Hippocampus [18, 19]. In this study, retrieval of a multiple number of spatiotemporal patterns has successfully been shown in spiking neural networks as in the manner similar to the present study of analog neural networks. These results on learning of multiple spatiotemporal patterns well explain the experimental result [9], in which not a single but multiple spatiotemporal patterns have been found to be memorized in the Hippocampus. Although the present study focuses only on learning of periodic patterns, the STDP-based learning scheme discussed here is applicable also to nonperiodic ones such as Poisson patterns [2, 19, 18, 56].

Interestingly reverse replay has been also recently simulated [46] in a model of the hippocampus with theta phase precession. In that model each unit is represented by a angular phase [20,55]. While the model [46] address the encoding and replay (forward and reversed) of only one specific pattern (i.e. the sequential activation of overlapping place fields at theta timescale due to phase precession), in our framework, we address the question of encode and replay multiple independent patterns (phase-coded memories), with multiple frequencies. The current study relates reverse replay to biologically plausible shape of STDP learning window. However, it still remains unclear if the STDP indeed has a contribution to occurrence of reverse replay. For example, Leibold et al. have discussed the possibility that short-term synapse plasticity causes time-compressed reverse replay of spatiotemporal pattern [57]. More efforts are necessary to reach a conclusion.

In our framework, the areas under the depression and potentiation part of the STDP learning window should balance, in order to maximize capacity. This balance between LTP and LTD is needed for the synaptic weights to respond to the asymmetric rather than the symmetric component of the unsubtracted cross-correlation. We found in [2] that storage capacity (the maximum of the number of patterns for successful retrieval) diverge in the thermodynamic limit when the ratio r of the area under the LTD part of the STDP window to the area under the LTP part is 1. The upper and lower bounds of r needed to store P patterns can be computed in our framework as functions of P and ω^μ [2]. If tight control is not maintained on the balance between synaptic potentiation and depression due to STDP, uncorrelated background firing will have a dominate effect.

Activity-dependent modification of synaptic strengths due to the proposed learning rule in eq. (3) is sensitive to correlations between pre- and post-synaptic firing over timescales of tens of ms when the range of $A(\tau)$ is tens of ms. The window $A(\tau)$ is the measure of the strength of synaptic change when there's a time delay τ between pre and post-synaptic activity. Writing eq. (3), implicitly we have assumed that the effects of separate spike pairs due to STDP sum linearly. However note that nonlinear effects have been observed when both pre- and post-synaptic neurons fire simultaneously at more then 40 Hz [49,50], therefore our model holds only in the case of lower firing rates, and in those case where linear summation is a good approximation. Also, we do not take into account here effects of saturation.

Finally, using learning rule in (3), we have both positive and negative connections (see also eq. (9)). Clearly, real neurons are or excitatory or inhibitory, and therefore the sign of J_{ij} should be positive if the pre-synaptic unit j is excitatory, and negative otherwise. As a remedy, one may use only excitatory units (i.e. only the non-negative connections are allowed) and add a global inhibition. Ongoing study shows that, even with only excitatory units, and a proper value of global inhibition, the network can encode and replay multiple oscillatory patterns. Indeed, in numerical simulations, we have found that, with all excitatory units, adding a proper value of global inhibition leads to results like those found earlier, where negative weights were permitted. We also remark that negative weights can also be simply implemented by inhibitory interneurons with very short membrane time constants.

References

1. Braitenberg, V., Caianiello, E.R., Lauria, F., Onesto, N.: Il nuovo cimento XI(2), 278–282 (1959)
2. Yoshioka, M., Scarpetta, S., Marinaro, M.: Phys. Rev. E. 75, 051917 (2007)
3. Buzsaki, G., Draguhn, A.: Science. 304, 1926–1929 (2004)
4. Gelperin, A.: The Journal of Neuroscience 26(6), 1663–1668 (2006)
5. Whittinghon Traub Trend Neurosc. 26, 676 (2003)
6. Wang in Encyclopedia of Cognitive Science (2003)
7. Llinas, R.: Science 242, 1654 (1988)
8. Hutcheon, B., Yarom, Y.: Trends Neurosci. 23, 216 (2000)
9. Nadasdy, Z., Hirase, H., Czurk, A., Csicsvari, J., Buzski, G.: J. Neurosci. 19, 9497 (1999)
10. Lee, A., Wilson, M.: Neuron. 36(6), 1183–1194 (2002)
11. Euston, D.R., McNaughton, T. M.: Science 318(5853), 1147–1150 (2007)
12. Foster, D.J., Wilson, M.A.: Nature 440(7084), 680–683 (2006)
13. Aoyagi, T., Kitano, K.: PRE 55, 7424 (1997)
14. Aoyagi, T.: PRL 74, 4075 (1995)
15. Kuzmina, M.G., Manykin, E.A., Surina, I.I.: In: Sandoval, F., Mira, J. (eds.) IWANN 1995. LNCS, vol. 930, pp. 246–251. Springer, Heidelberg (1995)
16. Hoppensteadt, F.C., Izhikevich, E.M.: Weakly Connected Neural Networks. Springer, New York (1997)
17. van Vreeswijk, C., Hansel, D.: Neural Computation 13, 959–992 (2001)
18. Yoshioka, M.: Phys. Rev. E 65, 011903 (2001)

19. Yoshioka, M.: Phys. Rev. E 66, 061913 (2002)
20. Yamaguchi, Y.: Biol. Cybern. 89, 1–9 (2003)
21. Arecchi, T.F.: Physica A 338, 218–237 (2004)
22. Borisyuk, R.: Hoppensteadt Biol. Cybern. 81, 359–371 (1999)
23. Borisyuk, R., Denham, M., Hoppensteadt, F., Kazanovich, Y., Vinogradova, O.: Network: Computation in Neural Systems 12(1), 1–20 (2001)
24. Marinaro, M., Scarpetta, S.: Neurocomputing 58-60C, 279–284 (2004)
25. Scarpetta, S., Zhaoping, L., Hertz, J.: Neural Computation 14(10), 2371–2396 (2002)
26. Scarpetta, S., Zhaoping, L., Hertz, J.: NIPS 2000. In: Leen, T., Dietterich, T., Tresp, V. (eds.), vol. 13. MIT Press, Cambridge (2001)
27. Magee, J.C., Johnston, D.: Science 275, 209–212 (1997)
28. Markram, H., Lubke, J., Frotscher, M., Sakmann, B.: Science 275, 213 (1997)
29. Nishiyama, M., Hong, K., Mikoshiba, K., Poo, M.-m., Kato, K.: NATURE 408, 584 (2000)
30. Bi, G.Q., Poo, M.M.: J. Neurosci. 18, 10464–10472 (1998)
31. Debanne, D., Gahwiler, B.H., Thompson, S.M.: J. Physiol. 507, 237–247 (1998)
32. Debanne, D., Gahwiler, B.H., Thompson, S.M.: Proc. Nat. Acad. Sci. 91, 1148–1152 (1994)
33. Feldman, D.E.: Neuron 27, 45–56 (2000)
34. Bi, G.Q., Poo, M.M.: Annual Review Neuroscience 24, 139–166 (2001)
35. Hasselmo, M.E.: Neural Computation 5, 32–44 (1993)
36. Hasselmo, M.E.: Trend in Cognitive Sciences 3(9), 351 (1999)
37. Gerstner, W., Kempter, R., van Leo, H.J., Wagner, H.: Nature 383 (1996)
38. Kempter, R., Gerstner, W., van Hemmen, L.: Physical Review E 59(5), 4498–4514 (1999)
39. Song, S., Miller, K.D., Abbott, L.F.: Competitive. Nat. Neurosci. 3(9), 919–926 (2000)
40. Rao, R.P., Sejnowski, T.J.: Neural Comput. 13(10), 2221–2237 (2001)
41. Lengyel, M., Kwag, J., Paulsen, O., Dayan, P.: Nature Neuroscience 8, 1677–1683 (2005)
42. Cook, J.: J. Phys. A 22, 2057 (1989)
43. Arenas, A., Perez Vicente, C.J.: Europhys. Lett. 26, 79 (1994)
44. Yoshioka, M., Shiino, M.: Phys. Rev. E 61, 4732 (2000)
45. Lengyel, M., Huhn, Z., Erdi, P.: Biol. Cybern. 92, 393–408 (2005)
46. Molter, C., Sato, N., Yamaguchi, Y.: Hippocampus 17, 201–209 (2007)
47. Li, Z., Hertz, J.: Network: Computation in Neural Systems 11, 83–102 (2000)
48. Drew, P.J., Abbott, L.F.: PNAS 2006 (2006)
49. Sjostrom, P.J., Turrigiano, G., Nelson, S.B.: Neuron 32, 1149–1164 (2001)
50. Froemke, R.C., Dan, Y.: Nature 416, 433–438 (2002)
51. Abarbanel, H., Huerta, R., Rabinovich, M.I.: PNAS 99(15), 10132–10137 (2002)
52. Harris, K.D., Csicsvari, J., Hirase, H., Dragoi, G., Buzsaki, G.: Nature 424, 552 (2003)
53. Kenet, T., Bibitchkov, D., Tsodyks, M., Grinvald, A., Arieli, A.: Nature 425, 954 (2003)
54. Fiser, J., Chiu, C., Weliky, M.: Nature 43, 573 (2004)
55. Hoppensteadt, F.C.: An introduction to the Mathematics of Neurons. Cambridge University Press, Cambridge (1986)
56. Yoshioka, M., Scarpetta, S., Marinaro, M.: In: de Sá, J.M., Alexandre, L.A., Duch, W., Mandic, D.P. (eds.) ICANN 2007. LNCS, vol. 4668, pp. 757–766. Springer, Heidelberg (2007)
57. Leibold, C., Gundlfinger, A., Schmidt, R., Thurley, K., Schmitz, D., Kempter, R.: Proc. Natl. Acad. Sci. 105, 4417–4422 (2008)

A Biophysical Model of Cortical Up and Down States: Excitatory-Inhibitory Balance and H-Current

Zaneta Navratilova and Jean-Marc Fellous

ARL Division of Neural Systems, Memory and Aging
University of Arizona, Tucson, AZ, USA

Abstract. During slow-wave sleep, cortical neurons oscillate between up and down states. Using a computational model of cortical neurons with realistic synaptic transmission, we determined that reverberation of activity in a small network of about 40 pyramidal cells could account for the properties of up states in vivo. We found that experimentally accessible quantities such as membrane potential fluctuations, firing rates and up state durations could be used as indicators of the size of the network undergoing the up state. We also show that the H-current, together with feed-forward inhibition can act as a gating mechanism for up state initiation.

1 Introduction

Slow wave sleep (SWS) is an active brain state in which memory consolidation and replay of neural activity patterns occur [1]. This stage of sleep is characterized by a slow (~0.5 Hz) oscillation in the cortical electroencephalogram (EEG). At the single cell level, cortical neurons switch between two states: an 'up state,' during which the membrane potential of the neurons is higher and the neurons spike frequently, and a 'down state' when the neurons are essentially silent [2, 3]. Although this oscillation can be highly synchronous between distant cortical regions (for example between the prefrontal and entorhinal cortex [4]), up states are also observed within cortical slices. *In vitro*, local glutamate application can initiate an up state in local neurons, causing a wave of up state onsets to spread across the slice [5]. In other slice preparations up states are more sporadic, but show repeating and ordered sequences of onsets [6]. Ordered up state onsets have also been observed *in vivo* [7]. Altogether, these data suggest that up states are local network phenomena that can be initiated by surrounding activity.

Up states recorded from different cortical regions do not have the same properties. For example, the firing rates of neurons may stay constant or decrease during the duration of the up state, depending on the cortical region under consideration (Andrea Hasenstaub, personal communication). Also, a precise mix of excitation and inhibition is necessary for the generation of up states, but the data on whether that mix includes more inhibition, excitation or a balanced amount is still unclear. Three groups have reported different inhibition to excitation ratios: 1:1 [8], 1:10 [9], or 2:1 [10]. The discrepancies can probably be attributed to differences in the cortical regions recorded, differences in the animal's species and differences in the induction and

M. Marinaro, S. Scarpetta, and Y. Yamaguchi (Eds.): Dynamic Brain, LNCS 5286, pp. 61–66, 2008.

depth of sleep (anesthesia or natural sleep). Because data from different preparations are so varied, it is likely that different network properties in different cortical regions lead to different properties of up states. We hypothesize here that one of the major differences leading to differences in up states is the size of the network being recruited.

How up states are generated *in vivo* is still unknown, but it is thought that a pulse of synchronous excitation from the hippocampus [11], thalamus [12] or other cortical neurons may be key. While network activity is certainly at play, intrinsic membrane properties may synergistically contribute as well [13]. We hypothesized that the H-current may play a major role in the initiation of up states, because it is a depolarizing current that activates during rapid changes in membrane potential (such as those caused by synchronous inputs) at below-threshold potentials. It has been shown to be involved in generating some types of rhythmic activity [14]. Additionally, this current is modulated by many types of neurotransmitters, such as those activated during SWS. For example, cortistatin, a peptide expressed in the cortex and hippocampus, increases the H-current and enhances slow wave sleep [15]. Also, dopamine enhances the amplitude and shifts the activation curve of this current [16]. Thus, we hypothesized that the H-current will enhance the likelihood of up state initiation.

2 Methods

We used the simulator NEURON to create a network model of biophysical neurons. Two types of neurons were simulated: excitatory, pyramidal-like neurons, and inhibitory GABAergic neurons. The excitatory neurons had a single somatic compartment, and a dendrite comprised of ten compartments. Passive leak currents adjusted to give an input resistance of 90 MΩ, were inserted in all compartments. Voltage-gated sodium and potassium currents were added to the soma [17] and adjusted to give an action potential generation threshold of -53mV. To control the bursting properties of pyramidal neurons, a calcium-activated potassium channel [18] and a calcium channel, pump and buffering [19] were added to the somatic compartment. In some simulations (see results, Fig. 3), an H-current was added to all compartments, comparable with experimental data [20]. Inhibitory neurons consisted of a single somatic compartment, and included voltage-gated sodium and potassium currents and passive leak currents adjusted to give an input resistance of 150 MΩ.

An Ornstein-Uhlenbeck background synaptic noise source [21] was added to the soma of each neuron to mimic the inputs from neurons outside of the simulated network, and was adjusted so that membrane potential fluctuations resembled those during a down state *in vivo*. Pyramidal neurons were connected to each other with AMPA/NMDA synapses showing facilitation and depression. These synapses were positioned onto a random dendritic compartment. There were approximately four times fewer inhibitory neurons than pyramidal neurons. Each inhibitory neuron received inputs from all the pyramidal neurons and output onto the somatic compartment of each pyramidal neuron to create shunting of the currents from the dendrite. These GABAergic synapses were deterministic [22]. Interneurons were not interconnected.

3   Results

A network of 26 pyramidal neurons and 6 inhibitory neurons was created as specified above. To generate an up state, a short (150ms) current pulse was given simultaneously to a few (~30%) model pyramidal neurons to mimic excitatory inputs from the thalamus, another cortical region, or the hippocampus. The conductances of the synaptic inputs were adjusted to obtain up state firing rates and pyramidal neuron membrane potential (V_m) averages and fluctuations (standard deviation) similar to those measured *in vivo* (Fig. 1 and Table 1). The V_m fluctuations were the only statistic that did not fit to measured data levels if this network were to generate up states, and so conductances were adjusted to make it as low as possible. The resulting up states terminated spontaneously after 500-2000 ms. Firing rates towards the end of the up state were constant until there was an abrupt end, indicating that activity did not just peter out.

Table 1. Comparison of up state statistics in model with *in vivo* data [4], [7], [8], [9], [10]

	26 excitatory neuron model	*In vivo* data
Excitatory neuron firing rates	10.4 +/-1.3 Hz	8-15 Hz
Inhibitory neuron firing rates	34.5 +/-5.3 Hz	15-30 Hz
Average up state membrane potential	-59.7 +/-1.9 mV	-50 to -60 mV
Up state membrane potential fluctuations	4.69 +/- 0.52 mV	2-3 mV
Average down state membrane potential	-68.3 +/- 0.5 mV	-65 to -75 mV
Down state membrane potential fluctuations	1.03 +/- 0.20 mV	0.6-2 mV
Duration of up state	1.168 +/- 0.470 s	0.4 – 1.6 s

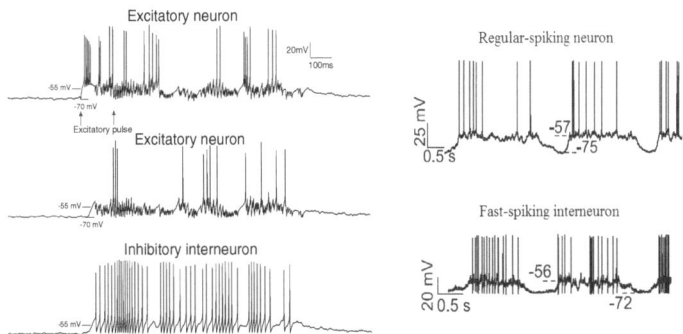

Fig. 1. Membrane potentials of three neurons (two pyramidal and one inhibitory) during an example up state generated by the model (*right*) compared to in vivo recordings of regular spiking and fast spiking neurons during the slow oscillation [8] (*left*)

To investigate how network size affected up state statistics, it was varied while keeping the proportion of excitatory and inhibitory neurons constant. Synaptic conductances were scaled proportionately to keep the overall synaptic inputs to each neuron approximately constant. This kept the average V_m of pyramidal neurons during up states constant (Fig. 2A), but changed the V_m fluctuations (Fig. 2B), which were mainly determined by the conductances of single synaptic events. The V_m fluctuations reached the measured *in vivo* level at a network size of 39 excitatory neurons, and appeared to asymptote within the *in vivo* measured range. The firing rates of both excitatory and inhibitory neurons during up states were larger in small networks (Fig. 2C). This may be due to the larger size of each individual synaptic conductance, which may allow the neuron to cross threshold more often even though the average input is approximately the same. The average V_m of inhibitory neurons, unlike that of pyramidal neurons, increased as network size decreased (Fig. 2A). This increase may be due to the slightly higher pyramidal neuron firing rates in the smaller networks, which cause a non-balanced increase in the number of excitatory inputs to the inhibitory neurons. Another statistic that changed with network size was up state duration, which increased for larger networks (Fig. 2D). These results show that a relatively small number of cells (39 pyramidal neurons, and 9 interneurons) can be recruited to generate and sustain up states comparable to those observed *in vivo*. Smaller networks required larger individual synaptic events than those observed *in vivo* to generate an up state.

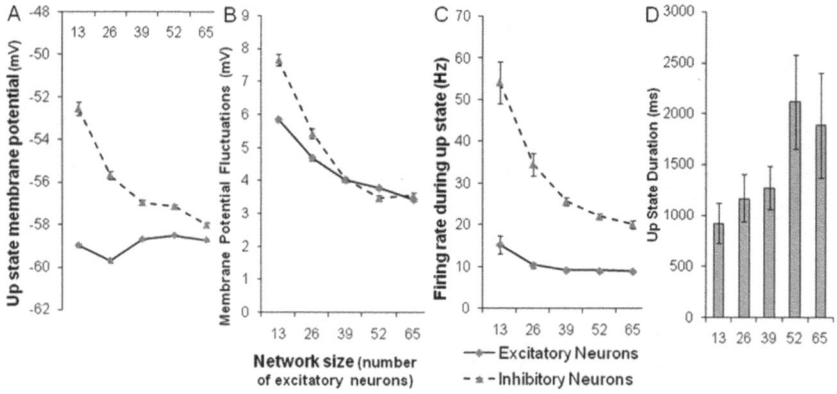

Fig. 2. The effects of changing network size on up state membrane potential (*A*), fluctuations (*B*), and firing rates (*C*) of both types of neurons in the network, and duration of the up state (*D*). *X-axis* is the number of pyramidal (excitatory) neurons in the network.

To test our hypothesis that intrinsic currents such as the H-current could contribute to the initiation of up states, we added this current to the pyramidal neurons in a 30-pyramidal neuron network. The H-current made up states more likely to be initiated by simultaneous excitatory inputs (Fig. 3, solid line compared to dotted line). Because the H-current activates at hyperpolarized membrane potentials, we reasoned that an inhibitory volley prior to excitatory inputs could further facilitate the elicitation of the up state. Such inhibition prior to excitation indeed enhanced the activation of the H-current and increased the probability of generating an up state (Fig. 3, dashed line). These differences in up state initiation were even bigger in a smaller network, where fewer simultaneously active neurons were needed to activate an up state (data not shown). The H-current did not have an effect on other properties of up states, such as their firing rates or duration.

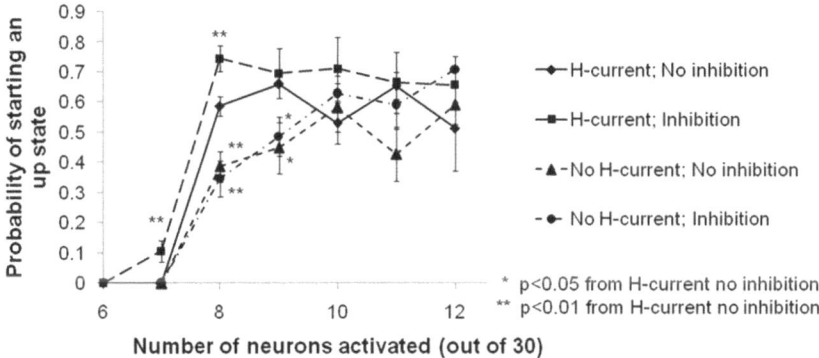

Fig. 3. The effects of the H-current in pyramidal neurons (*solid and dashed lines*), and an inhibitory volley prior to excitatory input (*dashed and dash-dotted lines*) on the probability of up state initiation. *Dotted line* is the control network, with no H-current or inhibitory volley.

4 Conclusions

Our model indicates that up states can be generated in a small network (as little as 40 neurons, if the network is fully connected) by the brief activation of a subset of those neurons. Activity reverberates in the network and spontaneously and abruptly shuts off, in the time scale seen *in vivo*. Also as seen *in vivo*, particular ratios of excitatory and inhibitory conductances were conducive to initiating up states.

In our model V_m fluctuation amplitudes, firing rates during the up state and up state duration all varied monotonically with network size. These experimentally measurable quantities can therefore be used as indicators of the size of the network (assuming full connectivity) that is responsible for an up state in preparation in which network size is not directly experimentally accessible. Thus, models such as this one can in principle be used to distinguish between different types of up states in different preparations and help determine the differences in their underlying mechanisms.

We further showed that adding an H-current increased the probability of generating an up state with the same number of synchronous inputs. Thus, the H-current may be one of the factors that enhance the initiation of up states. Feed-forward inhibition or hyperpolarization just prior to synchronous excitatory inputs made the likelihood of up state generation even greater in the presence of the H-current. This result suggests that feed-forward inhibition just prior to synchronous excitatory inputs could increase the likelihood of up state generation, and act as a gating mechanism. The neuromodulation of the H-current during SWS may be one of the factors that allows and/or enhances the initiation and gating of up states.

References

1. Sutherland, G.R., McNaughton, B.: Memory trace reactivation in hippocampal and neocortical neuronal ensembles. Curr. Opin. Neurobiol. 10, 180–186 (2000)
2. Steriade, M., Nuñez, A., Amzica, F.: A novel slow (< 1 Hz) oscillation of neocortical neurons in vivo: depolarizing and hyperpolarizing components. J. Neurosci. 13, 3252–3265 (1993)

3. Destexhe, A., Hughes, S.W., Rudolph, M., Crunelli, V.: Are corticothalamic 'up' states fragments of wakefulness? Trends Neurosci. 30, 334–342 (2007)
4. Isomura, Y., Sirota, A., Ozen, S., Montgomery, M., Mizuseki, K., Henze, D.A., Buzsaki, G.: Integration and Segregation of Activity in Entorhinal-Hippocampal Subregions by Neocortical Slow Oscillations. Neuron 52, 871–882 (2006)
5. Sanchez-Vives, M.V., McCormick, D.A.: Cellular and network mechanisms of rhythmic recurrent activity in neocortex. Nat. Neurosci. 3, 1027–1034 (2000)
6. Ikegaya, Y., Aaron, G., Cossart, R., Aronov, D., Lampl, I., Ferster, D., Yuste, R.: Synfire chains and cortical songs: temporal modules of cortical activity. Science 304, 559–564 (2004)
7. Luczak, A., Barthó, P., Marguet, S.L., Buzsáki, G., Harris, K.D.: Sequential structure of neocortical spontaneous activity in vivo. Proc. Natl. Acad. Sci. 104, 347–352 (2007)
8. Haider, B., Duque, A., Hasenstaub, A.R., McCormick, D.A.: Neocortical Network Activity In Vivo Is Generated through a Dynamic Balance of Excitation and Inhibition. J. Neurosci. 26, 4535–4545 (2006)
9. Waters, J., Helmchen, F.: Background synaptic activity is sparse in neocortex. J. Neurosci. 26, 8267–8277 (2006)
10. Rudolph, M., Pospischil, M., Timofeev, I., Destexhe, A.: Inhibition determines membrane potential dynamics and controls action potential generation in awake and sleeping cat cortex. J. Neurosci. 27, 5280–5290 (2007)
11. Battaglia, F.P., Sutherland, G.R., McNaughton, B.L.: Hippocampal sharp wave bursts coincide with neocortical "up-state" transitions. Learn. Mem. 11, 697–704 (2004)
12. Contreras, D., Steriade, M.: Cellular basis of EEG slow rhythms: a study of dynamic corticothalamic relationships. J. Neurosci. 15, 604–662 (1995)
13. Fellous, J.M., Sejnowski, T.J.: Regulation of persistent activity by background inhibition in an in vitro model of a cortical microcircuit. Cerebral Cortex 13, 1232–1241 (2003)
14. Luthi, A., McCormick, D.A.: H-Current: Properties of a Neuronal and Network Pacemaker. Neuron 21, 9–12 (1998)
15. Schweitzer, P., Madamba, S.G., Siggins, G.R.: The sleep-modulating peptide cortistatin augments the h-current in hippocampal neurons. J. Neurosci. 23, 10884–10891 (2003)
16. Chen, L., Yang, X.L.: Hyperpolarization-activated cation current is involved in modulation of the excitability of rat retinal ganglion cells by dopamine. Neuroscience 150, 299–308 (2007)
17. Golomb, D., Amitai, Y.: Propagating neuronal discharges in neocortical slices: computational and experimental study. J. Neurophys. 78, 1199–1211 (1997)
18. Destexhe, A., Contreras, D., Sejnowski, T.J., Steriade, M.: A model of spindle rhythmicity in the isolated thalamic reticular nucleus. J. Neurophysiol. 72, 803–818 (1994)
19. Destexhe, A., Babloyantz, A., Sejnowski, T.J.: Ionic mechanisms for intrinsic slow oscillations in thalamic relay neurons. Biophys. J. 65, 1538–1552 (1993)
20. Angelo, K., London, M., Christensen, S.R., Hausser, M.: Local and global effects of I(h) distribution in dendrites of mammalian neurons. J. Neurosci. 27, 8643–8653 (2007)
21. Destexhe, A., Rudolph, M., Fellous, J.M., Sejnowski, T.J.: Fluctuating synaptic conductances recreate in vivo-like activity in neocortical neurons. Neuroscience 107, 13–24 (2001)
22. Destexhe, A., Mainen, Z.F., Sejnowski, T.J.: Kinetic models of synaptic transmission. In: Koch, C., Segev, I. (eds.) Methods in Neuronal Modeling, 2nd edn. MIT press, Cambridge (1996)

Dynamical Architecture of the Mammalian Olfactory System

Leslie M. Kay

Department of Psychology,
Institute for Mind and Biology,
The University of Chicago,
Chicago, IL 60637,
USA
LKay@uchicago.edu

Abstract. The mammalian olfactory system shows many types of sensory and perceptual processing accompanied by oscillations at the level of the local field potential, and much is already known about the cellular and synaptic origins of these markers of coherent population activity. Complex, but chemotopic input patterns describe the qualitative similarity of odors, but animals can discriminate even very similar odorants. Coherent population activity signified by oscillations may assist the animals in discrimination of closely related odors. Manipulations to olfactory bulb centrifugal input and GABAergic circuitry can alter the degree of gamma (40-100 Hz) oscillatory coupling within the olfactory bulb, affecting animals' ability to discriminate highly overlapping odors. The demands of an odor discrimination can also enhance gamma oscillations, but this may depend on the cognitive demands of the task, with some tasks spreading the processing over many brain regions, accompanied by beta (15-30 Hz) instead of gamma oscillations.

Keywords: oscillation, gamma, beta, olfactory bulb, piriform cortex, hippocampus, synchrony, behavior.

1 Introduction

The textbook picture of sensory systems is a hierarchical one in which objective sensory input is passed to a primary sensory area, such as the thalamus or even primary sensory cortex, where some "preprocessing" happens. The processed signal is then transferred to the next cortical area where different streams of sensory information may be combined. At some level a "meaning" area adds to the signal, and the output then passes on to motor areas to produce a behavioral output. While this may seem an oversimplification to most neurophysiologists, this is the general framework that we teach our students and which biases our interpretations.

When we examine sensory processing in waking and intact animals we find that the sensory stimulus that arrives even at the first processing stage in the brain is already not objective. An individual seeks information in the environment with body movements and sensor array changes that influence what is assumed to be objective

M. Marinaro, S. Scarpetta, and Y. Yamaguchi (Eds.): Dynamic Brain, LNCS 5286, pp. 67–90, 2008.

sensory data. As a rat sniffs an odor the air in the nasal passages changes temperature and humidity. Sniffing dynamics may also direct the stimulus to optimal portions of the sensory epithelium (Schoenfeld and Cleland 2005), and head and body movements help an animal detect and locate an odor. Similar processes are at play in attentional behavior associated with any sensory stimulus. There are central mechanisms that have been discovered to accommodate for these search behaviors, such as the optokinetic effect in eye movements, but we still know relatively little about the dynamical interaction between brains and their environments.

Within the brain anatomical connections between sensory processing areas show dense bidirectional connectivity, and every sensory system shows strong effects of behavioral and arousal states that change the ways in which even very low level central neurons respond to stimulation. The olfactory system is no exception in this respect (Kay and Sherman 2007). This system also exhibits interesting dynamical behavior and has been studied from this perspective for many decades. This chapter will describe olfactory system behavioral physiology from a dynamical perspective, focusing on context-associated functional state and connectivity changes as exhibited primarily by neural oscillations.

2 Olfactory System Architecture

2.1 The Distributed Parts of the Olfactory System

The mammalian olfactory system is typically considered to be the collection of areas that receive monosynaptic input from the first cortical structure, the olfactory bulb[1]. This includes the anterior olfactory nucleus, the piriform cortex and portions of the entorhinal cortex and amygdala. However, bidirectional anatomical connections connect most parts of this system and connect the olfactory system with other systems, in particular the hippocampal system (Fig. 1).

Cortical and subcortical areas that project directly to the olfactory bulb include all parts of the olfactory system, plus areas such as the temporal pole of the hippocampus (van Groen and Wyss 1990), the septum, and the amygdala. In addition, almost every major neuromodulatory system sends output to the olfactory bulb; these include histaminergic input from the hypothalamus, serotonergic input from the Raphe nuclei, noradrenergic input from the locus coeruleus, and cholinergic input from the basal forebrain. It is commonly believed that olfactory bulb dopamine is entirely intrinsic, but in at least one species (sheep) there is a projection from the ventral tegmentum (Levy et al. 1999). Thus, the olfactory bulb receives perhaps more input from the brain than it does from the sensory receptors in the nose. All of this evidence points out that the hierarchical view of sensory processing is too simplistic, even at the anatomical level.

---

[1] Two reference works that give an excellent summary of olfactory system connectivity are Shepherd GM, and Greer CA. Olfactory bulb. In: The Synaptic Organization of the Brain, edited by Shepherd G. New York: Oxford University Press, 2003, p. 719. and Shipley MT, Ennis M, and Puche A. Olfactory System. In: The Rat Nervous System, edited by Paxinos G. San Diego: Academic Press, 2004.

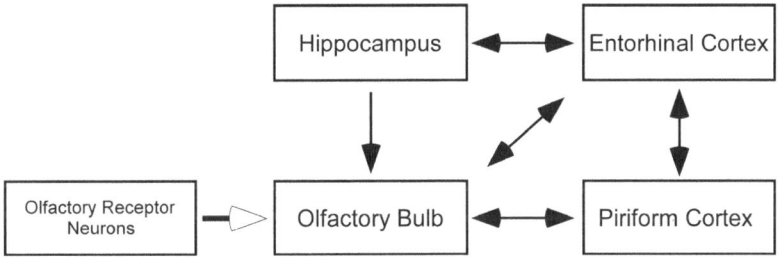

Fig. 1. The major bidirectional connections of the mammalian olfactory system. This figure illustrates that the system does not have a strictly feedforward architecture. The schematic represents brain regions and does not differentiate the different intraregional cell populations that form these connections. Not all bidirectional connections are shown.

2.2 Olfactory Bulb Input

The number of different functional olfactory receptor genes far exceeds the number of receptor types in other sensory systems, with functional genes numbering on the order of 1,000 in rodents and about 360 in humans (Buck and Axel 1991; Gilad et al. 2004; Glusman et al. 2001; Zhang and Firestein 2002). Sensory neurons in the nasal epithelium each express one type of olfactory receptor and project in an ordered fashion to the olfactory bulb. All of the sensory neurons expressing a given olfactory receptor type project their axons to a pair of identified glomeruli in the ipsilateral olfactory bulb (Fig 2) (Mombaerts et al. 1996). This makes the olfactory bulb input receptor-topic, although the logic of this ordering is not well understood. This ordered input translates to a relatively high-dimensional chemotopy. However, there is also no simple logic to the chemotopy, since individual receptors expressed on sensory receptor neurons are activated by parts of molecules, with optimal activation by a few mono-molecular odorants and lower levels of activation by other similar odorants. Since all molecules have multiple molecular features, a given odorant activates multiple receptor types. Thus, many odorants activate an individual glomerulus, and many glomeruli are activated by a single odorant. There is some stereotypy in monomolecular glomerular activation maps, but these maps appear to be very high dimensional (Leon and Johnson 2003). Furthermore, we have very little information regarding to which molecular features a given receptor may be most sensitive. To date, only a handful of receptors have been analyzed for optimal ligands, and only one receptor has been systematically analyzed (Araneda et al. 2000).

2.3 Basic Circuitry of the Olfactory Bulb

The olfactory bulb is a three layered paleocortical structure. The input layer is the glomerular layer, which contains principal neuron (mitral cell[2]) dendrites, glial cells and small juxtaglomerular cells which modulate sensory input and mitral cell activity (Fig 2). Juxtaglomerular cells are of many different types, but the most numerous are GABAergic, including a class of cells co-expressing GABA and dopamine. Sensory

---

[2] The principal neurons are actually mitral and tufted cells, which will be referred to as a group, mitral cells, in this article.

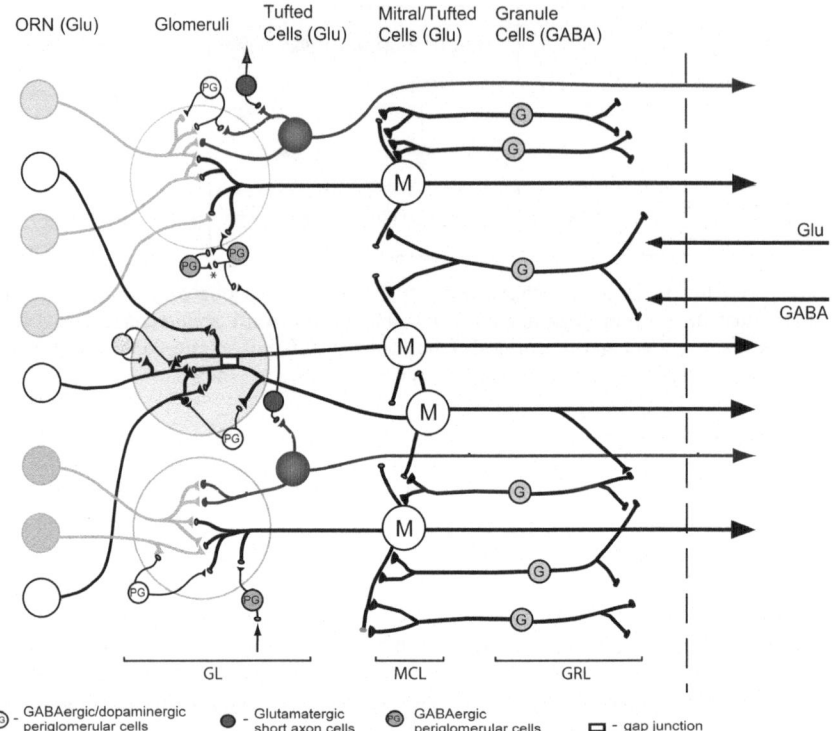

ORN (Glu)   Glomeruli   Tufted   Mitral/Tufted   Granule
                              Cells (Glu)   Cells (Glu)   Cells (GABA)

Glu

GABA

GL   MCL   GRL

⊙ - GABAergic/dopaminergic   ● - Glutamatergic   ⊚ GABAergic
    periglomerular cells   short axon cells   periglomerular cells   ▭ - gap junction

Fig. 2. Detailed schematic of the mammalian olfactory bulb circuit. Olfactory receptor neurons (ORNs) expressing the same odorant receptors send their axons to a common glomerulus in the olfactory bulb, where they synapse onto several types of cells. Shown are centrifugal GABA and glutamate projections synapsing onto lateral dendrites of granule cells, as discussed in the text. These centrifugal projections come from many brain areas and project specifically to the granular cell (GRL) and glomerular layers (GL). MCL: mitral cell layer. Dashed line indicates olfactory bulb boundary. Reprinted with permission from Kay LM, and Stopfer M. Seminars in Cell & Developmental Biology 17: 433-442, 2006.

neuron axons contact mitral cell dendrites directly in the glomeruli, and mitral cells typically have a single apical dendritic tuft in a single glomerulus, so they receive direct sensory input from only a single receptor type. (This is the only sensory system in which sensory neurons project directly to cortex, and it is one way in which substances can pass the blood-brain barrier.) Deeper layers include the mitral cell body layer about 700 μm below the pial surface, and deep to that layer is the relatively broad granule cell layer.

Mitral cells are glutamatergic and form reciprocal dendrodendritic synapses with GABAergic granule cells via their long lateral dendrites in a fiber-rich region between the glomeruli and the mitral cell layer, called the external plexiform layer. This synapse is functionally important, as it supports the gamma oscillations that are so prominent in the olfactory bulb (see Section 5.1). Because granule cells can release glutamate in a graded fashion in response to graded depolarization by mitral cells, inhibition of mitral cells can be graded, without the assistance of granule cell action potentials, or can be pulsatile when granule cells do fire. Slice studies show that physiological levels of Mg2+ are high enough that normal activation of granule cells by mitral cells at the dendrodendritic synapse is not enough to make the cell fire but could produce a graded release of GABA (Aroniadou-Anderjaska et al. 1999).

Centrifugal input to the olfactory bulb comes in primarily onto granule cells, making contact just below the mitral cell layer in the internal plexiform layer and throughout the granule cell layer. Glutamatergic input predominates from other cortical areas and activates granule cells on their basal dendrites. Because of the density of this input onto the granule cell population, it can provide strong modulation of olfactory bulb inhibition.

2.4 Connections between the Olfactory Bulb and Other Areas

Mitral cells project their long axons onto principal neurons in other parts of the olfactory system (Fig 2). Most prominent are the projections to the anterior olfactory nucleus and the anterior piriform cortex. These two areas also project more fibers back to the olfactory bulb than other olfactory structures. Other significant projections are to the olfactory tubercle, posterior piriform cortex, taenia tecta, indusium grisium, entorhinal cortex, amygdala, and the insula (Shipley and Adamek 1984). Projections to posterior piriform cortex are sparser than to the anterior part. Projections to the entorhinal cortex are onto the layer 2 stellate cells that project directly into the dentate gyrus of the hippocampus. This makes the pathway to the hippocampus from the olfactory bulb disynaptic through the entorhinal cortex. Feedback projections to the olfactory bulb from these areas are smaller than those from the anterior piriform and anterior olfactory nucleus. The medial entorhinal cortex receives olfactory output from the hippocampus, and this area also has projections back to the olfactory bulb granule cell layer (Biella and De Curtis 2000; Insausti et al. 1997). A more direct hippocampal projection to the olfactory bulb arises from the temporal pole of the hippocampus, where a subset of excitatory cells projects directly from CA1 to the granule cell layer of the olfactory bulb (Gulyas et al. 1998; van Groen and Wyss 1990). Many other areas project to the olfactory bulb, including the amygdala, taenia tecta, septum, and substantia nigra (Levy et al. 1999).

3 Olfactory System Electrophysiology

Olfactory system activity is studied at all levels of analysis, from the genetic basis of receptor projections to the olfactory bulb, to fMRI studies of meaning related odor processing. This review will focus on extracellular methods in anesthetized and waking animals and the various types of information they convey.

3.1 Signals and Analysis Tools

The local field potential (LFP) is generally considered to be the summed synaptic activity in a local population of neurons. This often gives a rough estimate of the summed spiking activity of local cells, but this is not always the case. Many cells do not fire as frequently as they receive substantial input. In the case of the olfactory bulb granule cells, this is significant. Thus, in the olfactory bulb, the LFP is a measure of the coherent activity of the local population, and as we will see below, this is a good estimate of the timing of mitral cell spikes. The scale of the measure is often referred to as "mesoscopic" to represent the middle level of analysis between single neurons and whole brain regions. The LFP is also a rough measure of what other brain regions receive from a given area, since the coherent activity is what survives down a pathway consisting of many axons with a distribution of conduction delays.

Analysis of LFP signals from single leads can be done with a variety of methods. Here we will concentrate on results from standard Fourier analysis. Other methods include wavelet and multitaper approaches. Both of these methods allow analysis of finer temporal structure in individual events, but all methods have weaknesses and strengths and should be chosen dependent on the question and the data. Coherence measures are often used to estimate cooperativity in a given frequency band (Chabaud et al. 1999; Fontanini and Bower 2005; Kay 2005; Kay and Freeman 1998; Lowry and Kay 2007; Martin et al. 2007). Phase estimates from these signals can be noisy and difficult to interpret and are only valid when coherence estimates are rather large. Furthermore, it is important to understand the source of the signals involved, because the absolute phase of a signal varies dependent on the position in the cortical tissue relative to the dipole field of the synaptic layer giving rise to the signal (Ferreyra Moyano et al. 1985; Freeman 1959; Martinez and Freeman 1984). To address questions of directionality in flow of these signals, some researchers are now using causal analysis methods (Bernasconi and Konig 1999; Brovelli et al. 2004; Seth 2005). These methods show great promise, particularly in a system with dense bidirectional connectivity and multiple generators of various rhythms.

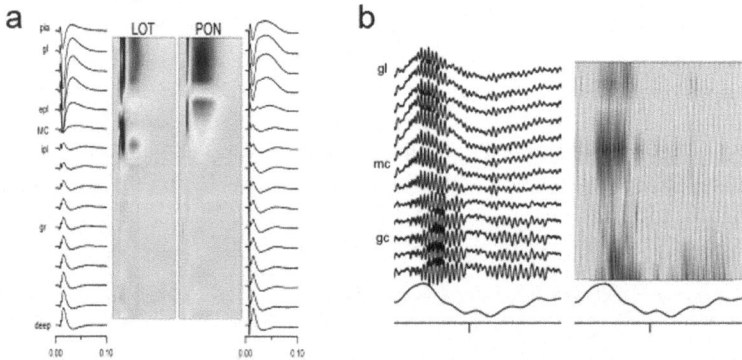

Fig. 3. Current source density profile from the olfactory bulb. **a)** Shock stimulus response in the olfactory bulb of an anesthetized rat (100 msec are shown). A shock to the lateral olfactory tract (LOT) stimulates the mitral cells antidromically and induces an oscillatory evoked potential with alternating current sinks and sources in the external plexiform layer. A shock stimulus to the primary olfactory nerve (PON) produces a similar effect, but the event begins in the more superficial glomerular layer and then activates the deeper layers of the bulb. **b)** Current source density profile in a waking rat exploring the cage. Alternating current source-sink pairs occur in the external plexiform layer and the granule cell layer. A deep source of lower frequency gamma 2 oscillation is evident in this event (Kay 2003). One second of data are shown.

Sources of oscillations can be investigated using several different methods. In waking animals a powerful tool for determining the synaptic sources of events associated with specific behavioral states is current source density (CSD). This method is a simple spatial derivative across the voltage signals recorded at successive depths perpendicular to the cortical layers. The data give a picture of the sources and sinks of current in the cortical tissue and are often displayed as color-coded spatiotemporal plots (Fig. 3a). An excellent reference paper which addresses the physics of estimating current and its sources in cortical tissue was published in 1985 (Mitzdorf 1985). This method has been used with some success in anesthetized animals (Martinez and Freeman 1984; Neville and Haberly 2003), but since many of the oscillations occur only in specific behavioral states, use of chronically implanted probes in waking animals is crucial to describing the sources of oscillations (Fig. 3b). CSD analysis has been used successfully in the hippocampus of waking rats to delineate the synaptic sources of some types of oscillations (Bragin et al. 1995; Kandel and Buzsaki 1997).

4 Coding Properties of the Input System

The anatomical structure of the input to the olfactory bulb predicts perceptual properties and neuron responsiveness to some extent. However, these properties depend in large part on an animal's behavioral state and prior experience with the stimuli.

The relatively structured input to the olfactory bulb provides some predictions about perceptual qualities of odorants. Anatomical projections from identified receptor types and glomerular activation maps are relatively stereotyped across individual animals, and this allows examination of activation pattern overlap for various odorants. Imaging studies show that odorants that are chemically similar activate overlapping regions of the glomerular sheet, while those that are dissimilar activate less overlapping regions (Grossman et al. in press; Johnson et al. 2004; Rubin and Katz 1999; Xu et al. 2003). If glomerular activation patterns represent a spatial code at the input level, then those odorants that have more overlapping patterns should smell more similar than those with less overlap, and mitral cells should respond to odorants similar to their "best" odorant.

4.1 Psychophysics

Psychophysical studies address the perceptual similarity of monomolecular odorants by several methods, and all of them rely on generalization of an odorant response to another test odorant using habituation or associative conditioning. An animal is trained to a single odorant, and probe odorants are used to test generalization of the learned response (Cleland et al. 2002). These tests verify that monomolecular odorants with larger overlap show more perceptual similarity than those with less overlap (Linster et al. 2001) (Fig. 5).

The above examples rely on generalization across a single dimension (e.g., chain length) for monomolecular stimuli. In reality, all natural odorants are blends or mixtures of many monomolecular odorants, and this makes studying them much more complicated. Even binary mixtures can have complex perceptual qualities, which have been roughly divided into two types. Elemental qualities are those in which the

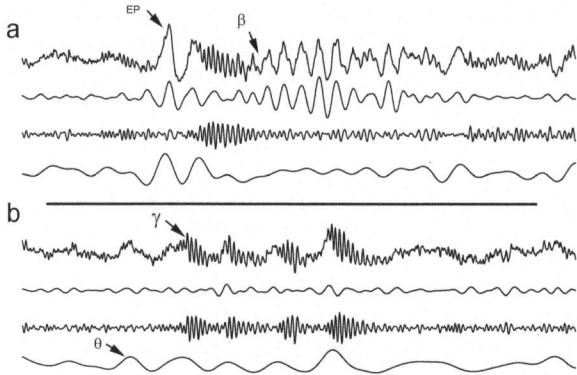

Fig. 4. Theta, beta and gamma oscillations from the olfactory bulb in waking rats. Each figure shows from top to bottom: raw data (1-475 Hz), beta band (15-35 Hz), gamma band (35-115 Hz), and theta band (1-12 Hz). 1.5 seconds of data are shown in each plot, both from the same rat in the same recording session. **a)** high amplitude beta band oscillation produced in response to odor sensitization. EP- sensory evoked potential; β- approximate beginning of beta oscillation. Note that the beta oscillation is preceded by a brief gamma frequency burst. **b)** gamma oscillations (marked by γ) associated with the transition from inhalation to exhalation during exploratory behavior. θ- marks respiratory wave in the theta band (inhalation is up). Note the relative absence of beta band activity during this episode.

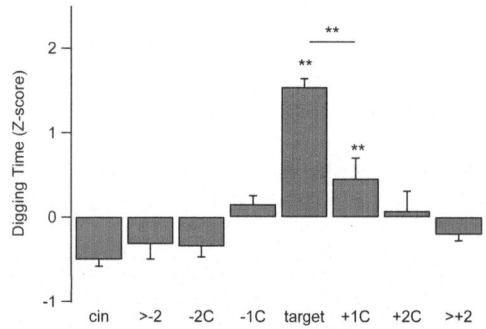

Fig. 5. Odor generalization along the dimension of carbon chain length. Mice were trained to respond by digging for an associated reward to a given aliphatic aldehyde on each day (randomly chosen 'target') over the course of ten trials. Digging times were measured in response to randomly ordered odorants in the test set, including the complete series of aldehydes (C3 – C10) and an unrelated odorant (cineole- cin). Digging times are presented as normalized Z-scores. Most mammals will generalize the association to the odorants nearest the trained odorant. The asymmetry in response is due to asymmetry in effective concentration due to differences in volatility across the odor set. There is a significant increase in response over the unrelated control (cin) for the trained aldehyde and the aldehyde one carbon longer ($p < 0.01$). There is also a significant difference between these two ($p < 0.01$).

binary mixture smells like the two components; configural (or synthetic) qualities are those in which the binary mixture smells like something different from the components. Most mixtures fall somewhere in between these two extremes, with one odorant overshadowing the other to greater or lesser extent (Kay et al. 2005). Other factors that may play a role are the physical properties of individual odorants and odorant concentrations (Kay et al. 2005; McNamara et al. 2007).

The generalization studies described above address the question "is the odorant you smell now, similar to the one which you smelled a few minutes ago?" Olfactory systems can do much more than find similarities; they are very good at finding differences. To see how the olfactory system finds differences, we turn to physiology.

4.2 Physiology

Relatively few studies have examined firing properties of olfactory bulb mitral cells in waking animals. This is because of the difficulty of isolating individual mitral cells due to their high background firing rates, packing density, and the very thin mitral cell layer. In waking rats and mice, most mitral cells show a significant modulation in firing rate associated with the respiratory cycle in which a burst of a few spikes is evoked upon inhalation (Bhalla and Bower 1997; Kay and Laurent 1999; Pager 1985). However, this is primarily during respiratory rates of less than 4 Hz. When rats transition to investigatory sniffing behavior, most mitral cells lose respiratory coupling. These two states are referred to as burst and tonic modes, representing inattentive and attentive states, respectively, by analogy with sensory thalamus neurons (Kay and Sherman 2007).

Mitral cells can respond to odor stimulation with an increase or decrease in firing rate in both anesthetized and waking animals (Bhalla and Bower 1997; Fletcher and Wilson 2003; Kay and Laurent 1999; Rinberg et al. 2006a). These neurons can also show changes in the temporal structure of firing relative to the LFP respiratory (theta) oscillation (Bhalla and Bower 1997; Pager 1983). In anesthetized rats, mice and rabbits, mitral cells show responses that suggest that they respond specifically to the chemotopic input, with nearby mitral cells responding to chemically similar odorants (Imamura et al. 1992; Katoh et al. 1993; Uchida et al. 2000). However, there is some dispute as to the generality of this receptive field response (Motokizawa 1996).

In waking rats and mice, the strongest modulation of firing rate is associated not with odor stimulation but rather the animal's behavioral state (waiting, sniffing, walking, etc.) (Fig. 6) (Kay and Laurent 1999; Rinberg et al. 2006a). These behavior-associated firing rate modulation patterns are specific to an individual neuron, are stable within a recording session and are similar across neurons recorded on the same electrode. Odor specificity, on the other hand, depends on the reward association of a particular odor. When the odor association is changed (e.g., from a positive (sucrose) to a negative (bitter taste) association) the odor selectivity of the neuron also changes (Kay and Laurent 1999). Even in anesthetized animals, a mitral cells' receptive field response can be altered by 'experience' (prolonged exposure to an odorant within the original receptive field) as has been shown for other sensory systems (Fletcher and Wilson 2003).

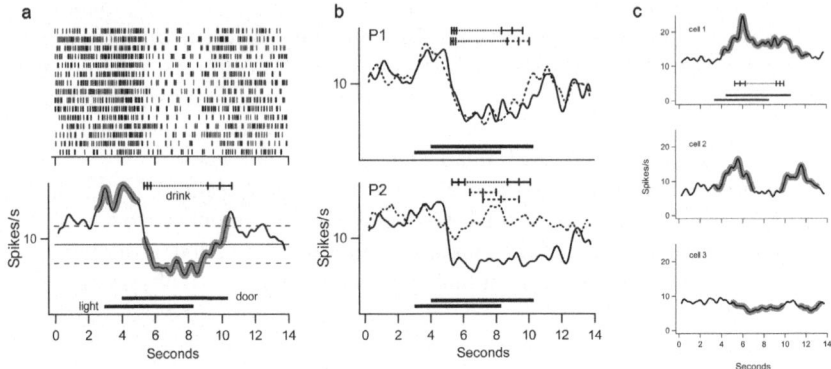

Fig. 6. Mitral cell responses are modulated by behavior. Mitral cell firing was recorded in a rat performing an odor association task: illumination of a house light signaled the opening of a door 1 second later. Behind the door was an odorant solution, which was either sweet or bitter. a) unit rasters and mitral cell firing histogram when only water and sucrose were present, with no additional odor exposure. Shading on histogram indicates time periods where firing was significantly different from the first 1 second of the trials. The horizontal line is the average firing rate before the onset of behavioral trials. Horizontal bars marked "light" and "door" signify the amount of time that the light was on and that the door was open. The bar marked "drink" shows the average onset and offset of the drinking response. b) the same neuron during subsequent odor trials. Top panel: both odors (trials randomly interleaved) were associated with a sweet solution. Bottom panel: one odor associated with the sweet solution and the other with a bitter solution. The rat ceases drinking the bitter solution in most trials (or drinks significantly later) after just a few learning trials. Note that the difference in firing rate histograms is restricted to the period in which the behavior is different. Odor selectivity responses (not shown) constitute a small but significant modulation on top of the behavioral modulation in about 10% of the neurons recorded. c) three simultaneously recorded cells that show similar background firing rates but different behavioral modulation patterns. (compiled and reprinted with permission from (Kay and Laurent 1999)).

5   Modulating the Input System

Olfactory bulb activity is strongly characterized by oscillations of the LFP, as described in section 3. The theta/respiratory oscillation is the most obvious and represents what the olfactory bulb sees of the inhalation/exhalation cycle; this also gives the neurophysiologist a tool by which to track gross olfactory behavior (sniffing rate and depth) in the LFP signal (see Fig. 4 for examples).

Initiating at the peak of inhalation is the olfactory bulb gamma oscillation, which is centered at about 70Hz in rats but can be as low as 40 Hz in cats and other larger mammals. This oscillation has been the focus of research in many laboratories since its discovery in the 1940s (Adrian 1942). Walter J. Freeman was the first to show that the gamma oscillation may play a functional role in perceptual processing. Freeman and colleagues recorded from arrays of 64 electrodes on the surface of the rabbit olfactory bulb coupled with conditioning and odor associations (Diprisco and Freeman 1985; Freeman and Schneider 1982). Several major findings resulted from these studies: 1) during odor sniffing, the frequency spectrum was dominated by power in the gamma band, except for the low frequency respiratory rhythm; 2) the waveform of the LFP was the same on all recording leads, as measured by RMS amplitude, or PCA or FFT decomposition; 3) the spatial pattern of amplitude of this common waveform was dependent on odor association, not on the odorant itself; 4) all spatial patterns associated with baseline and odor conditions changed when the association of any odor was changed. These last two results were reflected in single unit recordings more than a decade later (Kay and Laurent 1999).

5.1 Gamma Oscillation Circuit

Within an inhalation a given mitral cell may fire only 2 or 3 spikes, while the gamma oscillation itself may show 6 or more cycles. Thus, it is clear that the gamma oscillation in the LFP does not represent periodic firing of single neurons but rather is a population effect. However, the gamma oscillation is a very good indicator of how well the local mitral cells participate in this emergent population "synchrony" with the LFP. The gamma oscillation of the LFP has been shown to be a measure of the probability that a given mitral cell will fire during a gamma burst, with the phase of mitral cell firing being 90 degrees before the peak negativity of the gamma oscillation as measured at the pial surface (Eeckman and Freeman 1990). Because the LFP is the summed extracellular field from the neighboring neurons, as the firing of mitral cells near the recording electrode becomes more precise (closer to the -90 degree mark) the gamma oscillation should become larger.

The circuit that supports the gamma oscillation is the dendrodendritic synapse between mitral and granule cells (Fig 7). A similar effect is seen in piriform cortex (Freeman 1968). The reciprocal negative feedback circuit produces a cycle of excitation of granule cells by mitral cells, inhibition of mitral cells by granule cells, disexcitation of the granule cells, and finally disinhibition of the mitral cells. A similar sequence of events is seen with electrical stimulation of the olfactory tract, which produces an oscillatory evoked potential in both the olfactory bulb and piriform cortex. This effect was the subject of two early computational models (Freeman 1964; Rall and Shepherd 1968). Since both neurons participate in this event, both show strong coupling with the oscillation, but granule cells produce a more robust extracellular field, due to their parallel and bipolar geometry.

Current source density is a useful tool for finding the synaptic origins of oscillatory events in intact animals. By computing a spatial derivative across leads (or successively deeper penetrations) evenly arrayed perpendicular to the cell layers in a cortical structure, the sources and sinks of current flow can be estimated. Current source density studies on the olfactory bulbs of intact animals show that the oscillatory component of the shock stimulus evoked potential and the spontaneous gamma oscillation map onto the dendrodendritic synapse (Martinez and Freeman 1984; Neville and Haberly 2003).

Slice studies give us further insight into the circuitry involved in gamma oscillations. While precise spiking in granule cells can be elicited by patterned stimulation (Schoppa 2006b), it is not necessary for these cells to spike in order to provide inhibition to mitral cells and support very precise gamma-coupled spiking of mitral cells (Lagier et al. 2004). Current source density has also been used on olfactory bulb slices, and the results support those done in intact animals (Aroniadou-Anderjaska et al. 1999).

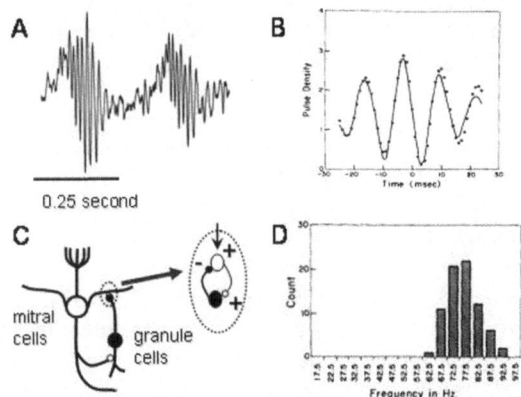

Fig. 7. Figure 7. Gamma oscillations are a network phenomenon. a) 500 msec of olfactory bulb LFP data showing two inhalations with gamma bursts. At this respiratory rate gamma bursts are common; with higher frequency sniffing, gamma bursts become irregular. **b)** The dendrodendritic synapse between mitral and granule cells appears to support the gamma oscillation, creating a local negative feedback circuit at the reciprocal synapse (Freeman 1975; Rall and Shepherd 1968). **c)** Pulse probability density (PPD) of mitral cell firing times relative to the peak of the gamma oscillation response. This is a measure of the probability of a single mitral cell firing, and the curve fit closely matches a gamma oscillation. **d)** Distribution of frequencies from many PPD curve fits. The range of frequencies matches the range of gamma oscillation frequencies. (a, c, d compiled with permission from Eeckman and Freeman 1990).

5.2 Manipulating the Circuit

Several types of manipulations to the olfactory bulb circuit result in changes to the power of gamma oscillations (Fig. 8). Removing centrifugal input to the olfactory bulb by various means results in enhancement of gamma oscillations in the olfactory bulb (Gray and Skinner 1988; Martin et al. 2004a; Martin et al. 2006). Gray and Skinner used a cryoprobe to temporarily cool the olfactory peduncle; they also recorded single unit mitral cell activity and showed that the locking to the gamma oscillation of individual mitral cells was more precise when centrifugal input was removed and gamma power was higher. This study also suggested that the gamma oscillation frequency was slightly decreased under blockade conditions. Martin and colleagues recorded olfactory bulb and piriform cortex activity during lidocaine blockade of only the feedback pathway, leaving the feedforward pathway intact (Fig. 8b). They showed that the enhanced gamma oscillations in the olfactory bulb were accompanied by somewhat enhanced gamma oscillations in the piriform cortex. Although that study did not examine the coherence of the oscillations in the two areas, other studies from intact animals show that gamma oscillations that occur simultaneously in these two highly interconnected structures can show very high levels of coherence (Kay and Freeman 1998; Lowry and Kay 2007).

Two manipulations of inhibition in this circuit produced opposite physiological and behavioral results. In the antennal lobe and mushroom body of many insects, gamma-like (20 Hz) oscillations are evoked by odorant stimulation of the antennae (Kay and

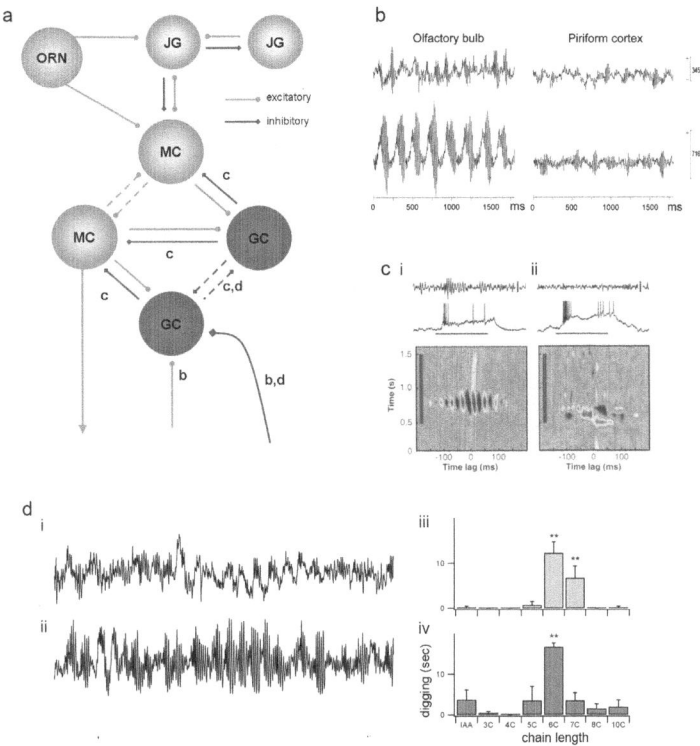

Fig. 8. Olfactory bulb circuit manipulations that affect population synchrony. a) Schematic of the olfactory bulb circuit (adapted with permission from (Freeman 1975), Academic Press). M- mitral cells, G- granule cells, JG- juxtaglomerular cells, ORN- olfactory receptor neurons. Deletions associated with manipulations in b-d are labeled with the respective letters on the schematic. **b)** Gamma activity in the olfactory bulb and piriform cortex under normal (top) and lidocaine blockade conditions (bottom). Gamma oscillations are enhanced in the olfactory bulb and piriform cortex when feedback to the olfactory bulb is blocked (reprinted with permission from (Martin et al. 2006)). **c)** odor induced 20 Hz oscillations in the locust antennal lobe (**i**: top trace and colored plot showing oscillatory correlation between unit and field responses during odor presentation). Oscillations are lost when picrotoxin, a GABA$_A$ receptor antagonist, is applied (**ii**: top trace and bottom colored plot). Middle plots- slow temporal structure of projection neuron firing is unchanged (reprinted with permission from (MacLeod and Laurent 1996)). **d)** Comparison between beta3 knockouts and littermate controls. **i)** Two seconds of LFP data the olfactory bulbs of control mice and knockouts (**ii**) during exploratory behavior. Note the obvious increase in gamma power. Control mice show normal generalization of a learned response to a similar odorant (**iii**), while the beta3 knockouts show no generalization (**iv**) (Nusser et al. 2001).

Stopfer 2006; Laurent et al. 1996). These oscillations are supported by the reciprocal excitation and inhibition between excitatory projection neurons and GABAergic inhibitory local neurons. The projection neurons fire approximately 90 degrees from the peak of the odor-evoked oscillation, like the mitral cells in the mammalian system. Application of picrotoxin, a GABAA receptor antagonist, to the antennal lobe removes inhibition in this circuit and abolishes odor-evoked oscillations but leaves the slower odorant responses in the spiking projection neurons intact (MacLeod and Laurent 1996) (Fig. 8c). Furthermore, downstream targets of these neurons lose their odor-selectivity when the oscillatory coupling in the antennal lobe circuit is disrupted (MacLeod et al. 1998). This same treatment applied to the antennal lobe in honeybees also disrupts oscillations, and it impairs the bees' discrimination of chemically similar but not dissimilar odorants (Stopfer et al. 1997). This is described as a deficit in "fine" odor discrimination, leaving "coarse" odor discrimination intact.

A different manipulation of GABAergic inhibition in mice produced opposite results (Nusser et al. 2001). Beta-3 knockout mice have a specific deletion of this subunit of the GABAA receptor. In the olfactory bulb this results in the specific ablation of functional GABAA receptors on granule cells, leaving other GABAergic inhibition in the bulb intact. This means that inhibition onto mitral cells at the reciprocal synapse is normal. The net effect on the circuit is to knock out mutual inhibition between granule cells and GABAergic drive to granule cells from other brain areas. This results in enhanced gamma oscillations in the olfactory bulb (Fig. 8d), and these mice are better than littermate control mice in discrimination of similar (fine) but not dissimilar (coarse) odors.

These two studies together suggest that gamma oscillation power, as a surrogate measure for neural firing precision, is related to discrimination of overlapping patterns. However, both of these treatments can cause severe disruption of other circuits, and the beta-3 knockout mice in particular have significant behavioral, anatomical and neurophysiological abnormalities (Homanics et al. 1997). What remained was proof that gamma oscillation power changes relative to the degree of overlap in the stimuli to be discriminated.

5.3 Changing the Intact Circuit with Context

It was left an open question whether or not animals can modify the amount of gamma oscillatory coupling on their own, without the aid of artificial manipulations. If the odorants to be discriminated have considerable overlap in their glomerular activation patterns (fine discrimination), more gamma oscillatory power should be seen in the olfactory bulb as compared to discriminating odorants with little overlap (coarse discrimination). In a study designed to test this hypothesis, Beshel and colleagues trained rats in a 2-alternative choice task to discriminate low or high overlap pairs of ketones and alcohols. In the case of low overlap, gamma oscillations showed a normal wide band irregular pattern during odor sniffing (Fig. 9a,b). When overlap was high and performance on the discrimination reached criterion levels, gamma oscillations were significantly enhanced (Fig. 9c).

Curiously, at the same time that olfactory bulb gamma oscillations were high, piriform cortex gamma oscillations were very low in power (Fig. 9d), contrary to what is seen during spontaneous exploratory behavior. In addition, olfactory bulb gamma oscillations did not maintain a high level once learning on an odor set reached criterion levels. Power was low at the beginning of each recording session and increased during the course of each session (Fig. 9d). These data suggest that it is not simply a wholesale increase in gamma power that accounts for enhanced odor discrimination ability, but that the role of gamma oscillatory precision in the olfactory bulb neural population is much more complex than previously thought.

The mechanism of this dynamic change in gamma band precision is still unknown, but the phenomena described above suggest one possible scenario. The timecourse of increase in gamma power in the olfactory bulb is similar to the dynamics of a sensory-evoked gamma power increase with application of muscarinic agonists on visual cortex in anesthetized cats (Rodriguez et al. 2004). The suppression of gamma oscillations in the piriform cortex also fits a cholinergic scenario. Modeling studies suggest that acetylcholine in the piriform cortex should reduce the influence of excitatory and inhibitory neurons within the piriform cortex, thus ablating piriform cortex

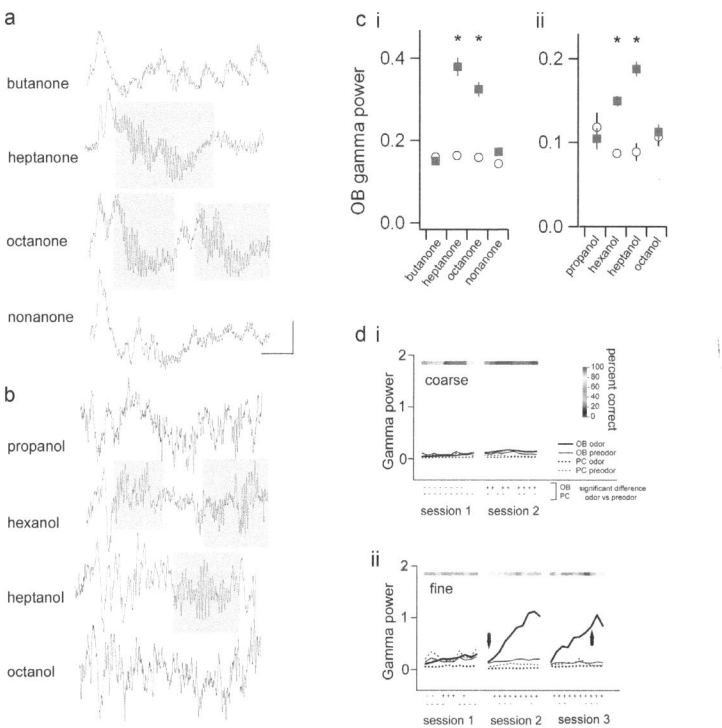

Fig. 9. Gamma oscillations are enhanced with task demand. a and **b**) Sample data for the odor sets used to test fine vs. coarse discrimination (the two center examples in **a- ketones** and **b- alcohols** are the fine odor discrimination pair, while the top and bottom in each group of 4 are the coarse pair). **c**) Olfactory bulb gamma band power distributions across 4 rats during odor sniffing for the odor sets in **a** (**c.i**) and **b** (**c.ii**). **d**) Evolution of gamma power through sessions with criterion performance throughout. **i**) Coarse discrimination shows no increase in gamma power in successive 10 trial blocks through the course of the experiment in either the olfactory bulb or piriform cortex. **ii**) Fine odor discrimination show a stead rise in gamma power only in the olfactory bulb during odor sniffing. Blockwise performance values are arrayed at the top in a color code. (compiled from (Beshel et al. 2007) and reprinted with permission).

gamma oscillations, and turning the pyramidal cells into simple input signal relays (Giocomo and Hasselmo 2007; Liljenstrom and Hasselmo 1995). The hypothesis receives some support from the increase in gamma oscillations under systemic administration of scopolamine, a cholinergic antagonist (Chabaud et al. 1999). This suggests that in the case of highly overlapping odor discrimination it is useful to maintain an 'objective' representation of the input pattern.

Other neuromodulators may also be able to modify olfactory bulb circuitry to increase or decrease gamma band cooperativity in the mitral cell population. In particular, D1 receptors at the reciprocal synapse could modify the effective inhibition from granule cells and decrease or increase precision accordingly (Brunig et al. 1999; Davila et al. 2003).

6   Multiple Functional Circuits

As described in section 3, the rat olfactory bulb also exhibits a prominent oscillation in the beta frequency band (15-30 Hz), which has been linked to odor learning, physico-chemical properties of odorants, and predator responses (Lowry and Kay 2007; Martin et al. 2007; Martin et al. 2004b; Zibrowski and Vanderwolf 1997). In anesthetized animals they are associated with the exhalation phase of the respiration and can show slightly different source-sink profiles than gamma oscillations (Buonviso et al. 2003; Neville and Haberly 2003). If gamma oscillations are the hallmark of odor associated activity in the olfactory bulb, what role do beta oscillations play? In particular, if beta oscillations are elicited upon learning odor discriminations, why aren't they evident in the task described above? This section will describe the conditions under which beta oscillations are observed, and detail the differences between beta and gamma oscillations and their behavioral correlates in the olfactory system.

6.1   Beta Oscillations and Learning

When rats learn an olfactory discrimination task in a Go/No-Go associative paradigm (response to one odor is rewarded, response to the other odor is penalized usually with a delay), enhancement of beta oscillations in the olfactory bulb and piriform cortex occurs simultaneous with the onset of correct performance in the discrimination (Martin et al. 2004b) (Fig. 10). Beta oscillations also occur in the hippocampus during odor sampling in this task, but the onset of power increase is not locked to the onset of correct performance for each odor set (Martin et al. 2007). Coherence between the olfactory bulb and hippocampus in the beta frequency band is enhanced as the rats learn to transfer the learned behavior to a new odor set, consistent with changes in the hippocampus that accompany olfactory rule learning (Zelcer et al. 2006). The coherence is maintained during odor sampling after this point. What is locked to correct performance is an enhanced coherence between the dorsal and ventral subfields of the hippocampus.

Beta oscillations involve a very large network and require intact feedback to the olfactory bulb (Martin et al. 2004a; Martin et al. 2006). When feedback to the olfactory bulb is ablated, the olfactory bulb produces only gamma oscillations during odor discrimination on the lesioned side, while the unlesioned side shows beta oscillations.

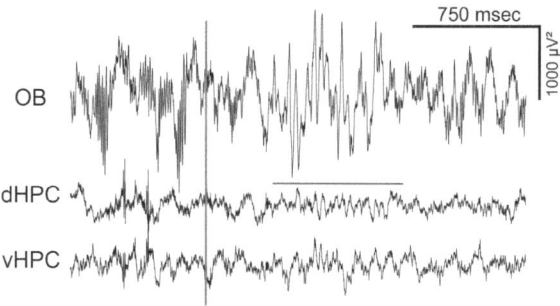

Fig. 10. Beta oscillations in the olfactory bulb are enhanced with learning in a Go/No-Go task. LFP data sample recorded simultaneously from the olfactory bulb (OB) and dorsal and ventral portions of the hippocampus (dHPC and vHPC, respectively). Blue vertical line signifies the time at which the odor was turned on, taking approximately 500 msec to arrive at the animal's nose. The horizontal red line indicates the beta oscillation. Reprinted from (Martin et al. 2007) with permission.

6.2 Olfactory Beta Oscillations in Other Conditions

Beta oscillations are also evoked using a sensitization paradigm in response to the predator odorants trimethyl thiazoline and 2-propylthietane (components of fox and weasel odors, respectively) and also to organic solvents such as toluene and xylene, absent any behavioral associative learning. The response to predator odorants resulted in the impression that these oscillations represented a specific predator response in rodents (Heale et al. 1994; Zibrowski and Vanderwolf 1997). However, we have recently argued that the volatility of these odorants may be the factor that induces beta oscillations (Lowry and Kay 2007). After 3-4 presentations of highly volatile organic chemicals (1-120 mm Hg theoretical vapor pressure), rats show prominent olfactory system beta oscillations restricted to the period of odor sniffing (Fig. 11). These oscillations arc not scen in anesthetized rats under the same exposure conditions.

6.3 Differences between Oscillations and Tasks

Why are there two different types of oscillations, each elicited during odor sniffing and each associated with learning an odor discrimination? One clue may lie in the tasks themselves. In section 5.3 I detailed the study that showed that gamma oscillations are enhanced during discrimination of highly overlapping odorants. In this study, a 2-alternative choice task was used in which both odors were rewarded upon responding by pressing a lever on one side or the other (one odor was paired with the right and one with the left lever). In the behavioral studies that produced beta oscillations (section 6.1), two different types of Go/No-Go tasks were used. Two of the three studies described in sections 5.3 and 6.1 were performed in the same operant chamber with the same odorant delivery system, the same odors and the same shaping protocols (Beshel et al. 2007; Martin et al. 2007). Both tasks use the same perceptual pathway but require different response associations, pointing out the cognitive difference between the two tasks, as has been shown in human studies (Braver et al. 2001). A difference in performance accuracy and ease is also seen between 2-alternative choice and Go/No-Go odor discrimination tasks (Abraham et al. 2004; Kay et al. 2006; Rinberg et al. 2006b; Uchida and Mainen 2003).

Go/No-Go tasks are typically easier for animals to learn and easier to transfer to new stimulus sets, and in our two studies this was indeed the case. The 2-alternative choice task requires an animal to respond with the same behavior in a different location to each of two stimuli. The Go/No-Go task requires a distinctly different behavior to be associated with each stimulus in a pair. There was little improvement in learning time from the first odor set to the last for the 2-alternative choice task (Beshel et al. 2007), while for the Go/No-Go task the number of trials required to reach criterion dropped significantly after the first training set (Martin et al. 2007).

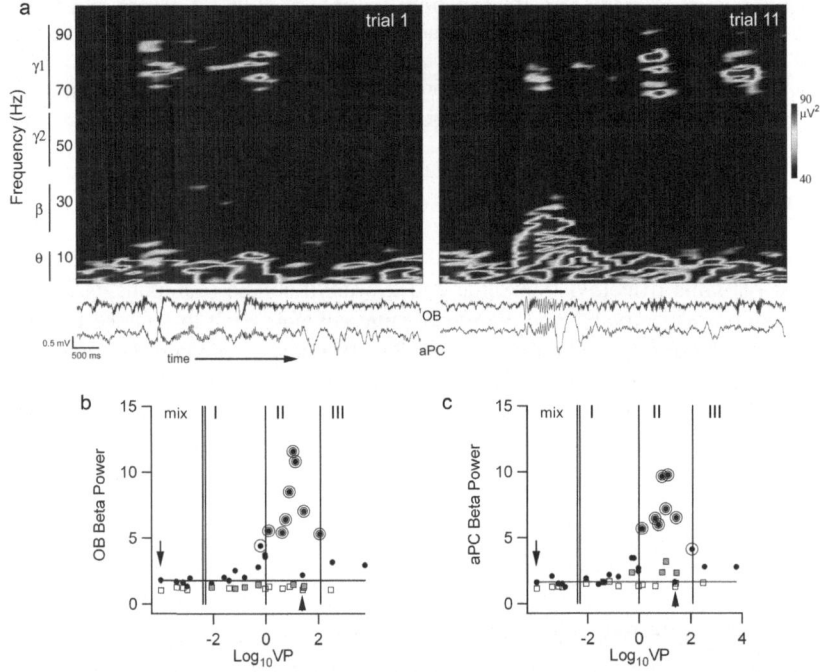

Fig. 11. Beta oscillations arise as a result of sensitization to highly volatile odorants. a) Dynamic power spectra from the olfactory bulb during a first (left) and 11th (right) odor presentation. LFP data from the olfactory bulb and piriform cortex are displayed below. The horizontal dark bar indicates the period during which the rat sniffed the odor swab. First trial investigation time is always significantly longer than subsequent trials. Note the change from gamma bursts to beta oscillations. b) Average olfactory bulb beta band power elicited in the olfactory bulb by odorants arrayed by volatility (theoretical vapor pressure) on a log scale. Ranges of theoretical vapor pressures are indicated (mixtures are on the far left) such that I corresponds to values below 1 mmHg, II to values between 1 and 120 mmHg, and III values above 120 mmHg. Circles around data points indicate significance values (single circle-p<0.05, double circle- p<0.01). c) Same as b but for piriform cortex. (Compiled and reprinted with permission from Lowry and Kay 2007).

What is different about the two systems required to produce the oscillations that is consistent with the differences in cognitive structure? The gamma oscillations produced in the 2-alternative choice task are local to the olfactory bulb, and the olfactory system appears to operate in a feed-forward fashion. The beta oscillations in the Go/No-Go task involve the entire extent of the olfactory and hippocampal systems and require a bidirectional connection between these structures and the olfactory bulb. This involvement of the downstream brain regions on primary olfactory processing may make the Go/No-Go task easier than the 2-alternative choice task. More research is needed to explain this difference.

7 Conclusion

The olfactory system presents fertile ground in which to study state-dependent dynamical neural structure. Many oscillatory states have been well-dissected, in particular the gamma oscillations, so that robust inferences can be made from population recordings. Decades of research from genetics, brain slices, and recordings in anesthetized, awake and behaving animals give us detailed information about the systems and subsystems involved in many characteristic events. New data showing changes in functional connectivity and oscillatory signature associated with task differences provide a means by which to more fully understand behavioral and cognitive influences on sensory dynamics. More research should be done to examine the sources of beta band oscillations in this system, to understand their role in learning and the relationship between learning (Martin et al. 2007) and odor sensitization (Lowry and Kay 2007). The dynamics of gamma oscillations during fine odor discrimination in the 2-alternative choice task suggest that it is not a simple binary effect that interacts with the sensory input overlap (Beshel et al. 2007). More research needs to be done to understand the mechanisms involved in the timecourse of gamma upregulation and its functional significance.

Acknowledgments

This chapter is the compilation of a set of lectures that I gave for the International School on Neural Networks, E. R. Caianiello course on The Dynamic Brain, at the Ettore Majorana Center in Erice, Italy in December 2007. I thank the organizers, Yoko Yamaguchi, Silvia Scarpetta and Maria Marinaro for the invitation, which led to many enlightening conversations. The work from my laboratory presented in this chapter is primarily from projects supported by NIDCD R01 DC007995.

References

1. Abraham, N.M., Spors, H., Carleton, A., Margrie, T.W., Kuner, T., Schaefer, A.T.: Maintaining accuracy at the expense of speed: Stimulus similarity defines odor discrimination time in mice. Neuron 44, 865–876 (2004)
2. Adrian, E.D.: The electrical activity of the mammalian olfactory bulb. Electroencephalography and Clinical Neurophysiology 2, 377–388 (1950)

3. Adrian, E.D.: Olfactory reactions in the brain of the hedgehog. Journal of Physiology 100, 459–473 (1942)
4. Araneda, R.C., Kini, A.D., Firestein, S.: The molecular receptive range of an odorant receptor. Nature Neuroscience 3, 1248–1255 (2000)
5. Aroniadou-Anderjaska, V., Ennis, M., Shipley, M.T.: Current-source density analysis in the rat olfactory bulb: laminar distribution of kainate/AMPA- and NMDA-receptor-mediated currents. Journal of Neurophysiology 81, 15–28 (1999)
6. Bernasconi, C., Konig, P.: On the directionality of cortical interactions studied by structural analysis of electrophysiological recordings. Biological Cybernetics 81, 199–210 (1999)
7. Beshel, J., Kopell, N., Kay, L.M.: Olfactory bulb gamma oscillations are enhanced with task demands. Journal of Neuroscience 27, 8358–8365 (2007)
8. Bhalla, U.S., Bower, J.M.: Multiday recordings from olfactory bulb neurons in awake freely moving rats: spatially and temporally organized variability in odorant response properties. Journal of Computational Neuroscience 4, 221–256 (1997)
9. Biella, G., De Curtis, M.: Olfactory inputs activate the medial entorhinal cortex via the hippocampus. Journal of Neurophysiology 83, 1924–1931 (2000)
10. Bragin, A., Jando, G., Nadasdy, Z., Hetke, J., Wise, K., Buzsaki, G.: Gamma (40-100 Hz) oscillation in the hippocampus of the behaving rat. Journal of Neuroscience 15, 47–60 (1995)
11. Braver, T.S., Barch, D.M., Gray, J.R., Molfese, D.L., Snyder, A.: Anterior cingulate cortex and response conflict: effects of frequency, inhibition and errors. Cerebral Cortex 11, 825–836 (2001)
12. Brovelli, A., Ding, M.Z., Ledberg, A., Chen, Y.H., Nakamura, R., Bressler, S.L.: Beta oscillations in a large-scale sensorimotor cortical network: Directional influences revealed by Granger causality. Proc. Natl. Acad. 101, 9849–9854 (2004)
13. Brunig, I., Sommer, M., Hatt, H., Bormann, J.: Dopamine receptor subtypes modulate olfactory bulb gamma-aminobutyric acid type A receptors. Proc. Natl. Acad. Sci. 96, 2456–2460 (1999)
14. Buck, L., Axel, R.: A novel multigene family may encode odorant receptors: a molecular basis for odor recognition. Cell 65, 175–187 (1991)
15. Buonviso, N., Amat, C., Litaudon, P., Roux, S., Royet, J.P., Farget, V., Sicard, G.: Rhythm sequence through the olfactory bulb layers during the time window of a respiratory cycle. European Journal of Neuroscience 17, 1811–1819 (2003)
16. Chabaud, P., Ravel, N., Wilson, D.A., Gervais, R.: Functional coupling in rat central olfactory pathways: a coherence analysis. Neuroscience Letters 276, 17–20 (1999)
17. Cleland, T.A., Morse, A., Yue, E.L., Linster, C.: Behavioral models of odor similarity. Behavioral Neuroscience 116, 222–231 (2002)
18. Davila, N.G., Blakemore, L.J., Trombley, P.Q.: Dopamine modulates synaptic transmission between rat olfactory bulb neurons in culture. Journal of Neurophysiology 90, 395–404 (2003)
19. Diprisco, G.V., Freeman, W.J.: Odor-related bulbar EEG spatial pattern-analysis during appetitive conditioning in rabbits. Behavioral Neuroscience 99, 964–978 (1985)
20. Eeckman, F.H., Freeman, W.J.: Correlations between unit firing and EEG in the rat olfactory system. Brain Research 528, 238–244 (1990)
21. Ferreyra Moyano, H., Cinelli, A.R., Molina, J.C.: Current generators and properties of early components evoked in rat olfactory cortex. Brain Research Bulletin 15, 237–248 (1985)

22. Fletcher, M.L., Wilson, D.A.: Olfactory bulb mitral-tufted cell plasticity: Odorant-specific tuning reflects previous odorant exposure. Journal of Neuroscience 23, 6946–6955 (2003)
23. Fontanini, A., Bower, J.M.: Variable coupling between olfactory system activity and respiration in ketamine/xylazine anesthetized rats. Journal of Neurophysiology 93, 3573–3581 (2005)
24. Freeman, W.J.: Distribution in time and space of prepyriform electrical activity. Journal of Neurophysiology 22, 644–665 (1959)
25. Freeman, W.J.: Linear distributed feedback model for prepyriform cortex. Experimental Neurology 10, 525–547 (1964)
26. Freeman, W.J.: Mass Action in the Nervous System, p. 489. Academic Press, New York (1975)
27. Freeman, W.J.: Relations between unit activity and evoked potentials in prepyriform cortex of cats. Journal of Neurophysiology 31, 337–348 (1968)
28. Freeman, W.J., Schneider, W.: Changes in spatial patterns of rabbit olfactory EEG with conditioning to odors. Psychophysiology 19, 44–56 (1982)
29. Gilad, Y., Wiebel, V., Przeworski, M., Lancet, D., Paabo, S.: Loss of olfactory receptor genes coincides with the acquisition of full trichromatic vision in primates. Plos. Biology 2, 120–125 (2004)
30. Giocomo, L.M., Hasselmo, M.E.: Neuromodulation by glutamate and acetylcholine can change circuit dynamics by regulating the relative influence of afferent input and excitatory feedback. Molecular Neurobiology 36, 184–200 (2007)
31. Glusman, G., Yanai, I., Rubin, I., Lancet, D.: The complete human olfactory subgenome. Genome Research 11, 685–702 (2001)
32. Gray, C.M., Skinner, J.E.: Centrifugal regulation of neuronal activity in the olfactory bulb of the waking rabbit as revealed by reversible cryogenic blockade. Experimental Brain Research 69, 378–386 (1988)
33. Grossman, K.J., Mallik, A.K., Ross, J., Kay, L.M., Issa, N.P.: Glomerular activation patterns and the perception of odor mixtures. European Journal of Neuroscience 27(10), 2676–2685 (2008)
34. Gulyas, A.I., Toth, K., McBain, C.J., Freund, T.F.: Stratum radiatum giant cells: a type of principal cell in the rat hippocampus. European Journal of Neuroscience 10, 3813–3822 (1998)
35. Heale, V.R., Vanderwolf, C.H., Kavaliers, M.: Components of weasel and fox odors elicit fast wave bursts in the dentate gyrus of rats. Behavioural Brain Research 63, 159–165 (1994)
36. Homanics, G.E., DeLorey, T.M., Firestone, L.L., Quinlan, J.J., Handforth, A., Harrison, N.L., Krasowski, M.D., Rick, C.E., Korpi, E.R., Makela, R., Brilliant, M.H., Hagiwara, N., Ferguson, C., Snyder, K., Olsen, R.W.: Mice devoid of gamma-aminobutyrate type A receptor beta3 subunit have epilepsy, cleft palate, and hypersensitive behavior. Proc. Natl. Acad. Sci. 94, 4143–4148 (1997)
37. Imamura, K., Mataga, N., Mori, K.: Coding of odor molecules by mitral/tufted cells in rabbit olfactory bulb. I. Aliphatic compounds. Journal of Neurophysiology 68, 1986–2002 (1992)
38. Insausti, R., Herrero, M.T., Witter, M.P.: Entorhinal cortex of the rat: Cytoarchitectonic subdivisions and the origin and distribution of cortical efferents. Hippocampus 7, 146–183 (1997)
39. Johnson, B.A., Farahbod, H., Xu, Z., Saber, S., Leon, M.: Local and global chemotopic organization: General features of the glomerular representations of aliphatic odorants differing in carbon number. Journal of Comparative Neurology 480, 234–249 (2004)

40. Kandel, A., Buzsaki, G.: Cellular-synaptic generation of sleep spindles, spike-and-wave discharges, and evoked thalamocortical responses in the neocortex of the rat. Journal of Neuroscience 17, 6783–6797 (1997)
41. Katoh, K., Koshimoto, H., Tani, A., Mori, K.: Coding of odor molecules by mitral/tufted cells in rabbit olfactory bulb. II. Aromatic compounds. Journal of Neurophysiology 70, 2161–2175 (1993)
42. Kay, L.M.: Theta oscillations and sensorimotor performance. Proceedings of the National Academy of Sciences 102, 3863–3868 (2005)
43. Kay, L.M.: Two species of gamma oscillations in the olfactory bulb: dependence on behavioral state and synaptic interactions. Journal of Integrative Neuroscience 2, 31–44 (2003)
44. Kay, L. M., Beshel, J., Martin, C.: When good enough is best. Neuron 51, 277–278 (2006)
45. Kay, L.M., Crk, T., Thorngate, J.: A redefinition of odor mixture quality. Behavioral Neuroscience 119, 726–733 (2005)
46. Kay, L.M., Freeman, W.J.: Bidirectional processing in the olfactory-limbic axis during olfactory behavior. Behavioral Neuroscience 112, 541–553 (1998)
47. Kay, L.M., Laurent, G.: Odor- and context-dependent modulation of mitral cell activity in behaving rats. Nature Neuroscience 2, 1003–1009 (1999)
48. Kay, L.M., Sherman, S.M.: An argument for an olfactory thalamus. Trends in Neurosciences 30, 47–53 (2007)
49. Kay, L.M., Stopfer, M.: Information processing in the olfactory systems of insects and vertebrates. Seminars in Cell & Developmental Biology 17, 433–442 (2006)
50. Lagier, S., Carleton, A., Lledo, P.M.: Interplay between local GABAergic interneurons and relay neurons generates gamma oscillations in the rat olfactory bulb. Journal of Neuroscience 24, 4382–4392 (2004)
51. Laurent, G., Wehr, M., MacLeod, K., Stopfer, M., Leitch, B., Davidowitz, H.: Dynamic encoding of odors with oscillating neuronal assemblies in the locust brain. Biological Bulletin 191, 70–75 (1996)
52. Leon, M., Johnson, B.A.: Olfactory coding in the mammalian olfactory bulb. Brain Research Reviews 42, 23–32 (2003)
53. Levy, F., Meurisse, M., Ferreira, G., Thibault, J., Tillet, Y.: Afferents to the rostral olfactory bulb in sheep with special emphasis on the cholinergic, noradrenergic and serotonergic connections. Journal of Chemical Neuroanatomy 16, 245–263 (1999)
54. Liljenstrom, H., Hasselmo, M.E.: Cholinergic modulation of cortical oscillatory dynamics. Journal of Neurophysiology 74, 288–297 (1995)
55. Linster, C., Johnson, B.A., Yue, E., Morse, A., Xu, Z., Hingco, E.E., Choi, Y., Choi, M., Messiha, A., Leon, M.: Perceptual correlates of neural representations evoked by odorant enantiomers. Journal of Neuroscience 21, 9837–9843 (2001)
56. Lowry, C.A., Kay, L.M.: Chemical factors determine olfactory system beta oscillations in waking rats. Journal of Neurophysiology 98, 394–404 (2007)
57. MacLeod, K., Backer, A., Laurent, G.: Who reads temporal information contained across synchronized and oscillatory spike trains? Nature 395, 693–698 (1998)
58. MacLeod, K., Laurent, G.: Distinct mechanisms for synchronization and temporal patterning of odor-encoding neural assemblies. Science 274, 976–979 (1996)
59. Martin, C., Beshel, J., Kay, L.M.: An olfacto-hippocampal network is dynamically involved in odor-discrimination learning. Journal of Neurophysiology 98, 2196–2205 (2007)
60. Martin, C., Gervais, R., Chabaud, P., Messaoudi, B., Ravel, N.: Learning-induced modulation of oscillatory activities in the mammalian olfactory system: The role of the centrifugal fibres. Journal of Physiology-Paris 98, 467–478 (2004a)

61. Martin, C., Gervais, R., Hugues, E., Messaoudi, B., Ravel, N.: Learning modulation of odor-induced oscillatory responses in the rat olfactory bulb: A correlate of odor recognition? Journal of Neuroscience 24, 389–397 (2004b)

62. Martin, C., Gervais, R., Messaoudi, B., Ravel, N.: Learning-induced oscillatory activities correlated to odour recognition: a network activity. European Journal of Neuroscience 23, 1801–1810 (2006)

63. Martinez, D.P., Freeman, W.J.: Periglomerular cell action on mitral cells in olfactory bulb shown by current source density analysis. Brain Research 308, 223–233 (1984)

64. McNamara, A.M., Magidson, P.D., Linster, C.: Binary mixture perception is affected by concentration of odor components. Behavioral Neuroscience 121, 1132–1136 (2007)

65. Mitzdorf, U.: Current source-density method and application in cat cerebral cortex: investigation of evoked potentials and EEG phenomena. Physiological Reviews 65, 37–100 (1985)

66. Mombaerts, P., Wang, F., Dulac, C., Chao, S.K., Nemes, A., Mendelsohn, M., Edmondson, J., Axel, R.: Visualizing an olfactory sensory map. Cell 87, 675–686 (1996)

67. Motokizawa, F.: Odor representation and discrimination in mitral/tufted cells of the rat olfactory bulb. Experimental Brain Research 112, 24–34 (1996)

68. Neville, K.R., Haberly, L.B.: Beta and gamma oscillations in the olfactory system of the urethane-anesthetized rat. Journal of Neurophysiology 90, 3921–3930 (2003)

69. Nusser, Z., Kay, L.M., Laurent, G., Homanics, G.E., Mody, I.: Disruption of GABA(A) receptors on GABAergic interneurons leads to increased oscillatory power in the olfactory bulb network. Journal of Neurophysiology 86, 2823–2833 (2001)

70. Pager, J.: Respiration and olfactory bulb unit activity in the unrestrained rat: statements and reappraisals. Behavioural Brain Research 16, 81–94 (1985)

71. Pager, J.: Unit responses changing with behavioral outcome in the olfactory bulb of unrestrained rats. Brain Research 289, 87–98 (1983)

72. Rall, W., Shepherd, G.M.: Theoretical reconstruction of field potentials and dendrodendritic synaptic interactions in olfactory bulb. Journal of Neurophysiology 31, 884–915 (1968)

73. Rinberg, D., Koulakov, A., Gelperin, A.: Sparse odor coding in awake behaving mice. Journal of Neuroscience 26, 8857–8865 (2006a)

74. Rinberg, D., Koulakov, A., Gelperin, A.: Speed-accuracy tradeoff in olfaction. Neuron 51, 351–358 (2006b)

75. Rodriguez, R., Kallenbach, U., Singer, W., Munk, M.H.J.: Short- and long-term effects of cholinergic modulation on gamma oscillations and response synchronization in the visual cortex. Journal of Neuroscience 24, 10369–10378 (2004)

76. Rubin, B.D., Katz, L.C.: Optical imaging of odorant representations in the mammalian olfactory bulb. Neuron 23, 499–511 (1999)

77. Schoenfeld, T.A., Cleland, T.A.: The anatomical logic of smell. Trends in Neurosciences 28, 620–627 (2005)

78. Schoppa, N.E.: AMPA/Kainate receptors drive rapid output and precise synchrony in olfactory bulb granule cells. Journal of Neuroscience 26, 12996–13006 (2006a)

79. Schoppa, N.E.: Synchronization of olfactory bulb mitral cells by precisely timed inhibitory inputs. Neuron 49, 271–283 (2006b)

80. Seth, A.K.: Causal connectivity of evolved neural networks during behavior. Network-Computation in Neural Systems 16, 35–54 (2005)

81. Shepherd, G.M., Greer, C.A.: Olfactory bulb. In: Shepherd, G. (ed.) The Synaptic Organization of the Brain, p. 719. Oxford University Press, New York (2003)

82. Shipley, M.T., Adamek, G.D.: The connections of the mouse olfactory bulb: a study using orthograde and retrograde transport of wheat germ agglutinin conjugated to horseradish peroxidase. Brain Research Bulletin 12, 669–688 (1984)
83. Shipley, M.T., Ennis, M., Puche, A.: Olfactory System. In: Paxinos, G. (ed.) The Rat Nervous System. Academic Press, San Diego (2004)
84. Stopfer, M., Bhagavan, S., Smith, B.H., Laurent, G.: Impaired odour discrimination on de-synchronization of odour- encoding neural assemblies. Nature 390, 70–74 (1997)
85. Uchida, N., Mainen, Z.F.: Speed and accuracy of olfactory discrimination in the rat. Nature Neuroscience 6, 1224–1229 (2003)
86. Uchida, N., Takahashi, Y.K., Tanifuji, M., Mori, K.: Odor maps in the mammalian olfactory bulb: domain organization and odorant structural features. Nature Neuroscience 3, 1035–1043 (2000)
87. van Groen, T., Wyss, J.M.: Extrinsic projections from area CA1 of the rat hippocampus: olfactory, cortical, subcortical, and bilateral hippocampal formation projections. Journal of Comparative Neurology 302, 515–528 (1990)
88. Xu, F., Liu, N., Kida, I., Rothman, D.L., Hyder, F., Shepherd, G.M.: Odor maps of alde-hydes and esters revealed by functional MRI in the glomerular layer of the mouse olfactory bulb. Proc. Natl. Acad. Sci. 100, 11029–11034 (2003)
89. Zelcer, I., Cohen, H., Richter-Levin, G., Lebiosn, T., Grossberger, T., Barkai, E.: A cellu-lar correlate of learning-induced metaplasticity in the hippocampus. Cerebral Cortex 16, 460–468 (2006)
90. Zhang, X., Firestein, S.: The olfactory receptor gene superfamily of the mouse. Nature Neuroscience 5, 124–133 (2002)
91. Zibrowski, E.M., Vanderwolf, C.H.: Oscillatory fast wave activity in the rat pyriform cor-tex: relations to olfaction and behavior. Brain Research 766, 39–49 (1997)

From Behaviour to Brain Dynamics

Allen Cheung

Queensland Brain Institute
School of Information Technology and Electrical Engineering,
The University of Queensland, QLD 4072, Australia
acheung@itee.uq.edu.au

Abstract. It is well accepted that medium to long range navigation requires the use of an external directional reference i.e. a compass. Cheung et al (2007) recently demonstrated through theory and simulation the quantitative significance of the compass. It was shown that navigating agents using and not using a compass could be differentiated on the basis of the population behaviour. In the current work, theory and simulation results will be presented on ways to characterize individual paths on the basis of whether the system was using an external directional reference. Thus it is demonstrated that important information concerning the neural input used by a navigating animal may be inferred probabilistically from its behaviour.

1   Population-Based Behaviour

It is an age old problem to understand the behaviour of a control system e.g. the brain, based on its input and output, and yet that is often the challenge facing a neuroethologist. One particular behavioural output which is of interest is animal navigation. Current technology is capable of tracking with unprecedented precision the position and orientation of animals. However, it is only in strict laboratory conditions that such behaviour may be measured concurrently with neurodynamics. It has been shown that the behavioural characteristics of a population of navigating agents differed quantitatively and qualitatively depending on whether a compass (allothetic directional cue) is available to the population (Cheung et al 2007). In particular, the expected displacement of an agent has a finite upper limit if it did not use a compass (using only idiothetic cues)! This is accompanied by a positional uncertainty which asymptotically increases more rapidly than that of a Pearson's random walk (Pearson 1905). Using a compass, however, an agent could travel arbitrarily far along any predefined direction (axis of intended locomotion), with relatively small positional uncertainty. With such a dichotomy in population behaviour, it stands to reason that if it is feasible to obtain population estimates such as the positional mean and variance along and perpendicular to the axis of intended locomotion, then it should be possible to determine with some confidence what class of directional sensory cue was used (idiothetic or allothetic). The expected properties are summarized in Table 1.

M. Marinaro, S. Scarpetta, and Y. Yamaguchi (Eds.): Dynamic Brain, LNCS 5286, pp. 91–95, 2008.
© Springer-Verlag Berlin Heidelberg 2008

Table 1. Population-based characteristics of different forms of directed walks

Type of Directed Walk	
Idiothetic	*Allothetic*
$\overline{x_{total}} < E_{max} < n\overline{x_1}$	$\overline{x_{total}} = n\overline{x_1}$
$s^2_{total} > n^2 > ns^2_1$	$n^2 > s^2_{total} = ns^2_1$
$\lim_{n\to\infty} s^2_x(n) = s^2_y(n)$	$\lim_{n\to\infty} s^2_x(n) \neq s^2_y(n)$

The average displacement after n steps along the axis of intended locomotion is denoted $\overline{x_{total}}$, while s^2_{total} denotes the sample variance in position along any particular axis. The Y direction is designated as being perpendicular to the axis of intended locomotion. The asymptotic limit for $\overline{x_{total}}$ during a simple idiothetic directed walk is denoted E_{max} and was shown by Cheung et al (2007) to be

$$E_{max} = \lim_{n\to\infty} \langle X_{total} \rangle = \mu_L \frac{\beta}{1 - \beta} \tag{1}$$

where $\beta = \langle \cos \Delta \rangle$. These characteristics allow populations of navigating agents undergoing IDWs or ADWs to be differentiated.

2   Individual-Based Behaviour

Despite the analytical rigour of the population-based results, it may be difficult in practice to obtain a sufficient sample size to be confident of the estimates of population parameters. Furthermore, repeated trials can only be pooled with confidence if the axis of intended locomotion is known for each trial and therefore aligned.

A very different approach is currently being developed to quantify the IDW vs ADW character of an individual directed walk (see Fig 1), without a priori knowledge of the axis of intended locomotion, magnitude of random errors, or existence of bias. The geometric construct is analogous to the simple directed walks presented in Cheung et al (2007). The angular error at step t is denoted Δ_t. Hence for a simple idiothetic directed walk, the allocentric heading Θ_t following t steps is the sum of all preceding Δ's. In contrast, for the ideal allothetic directed walk, a compass is used to reset heading errors at each step such that by the $t'th$ step, errors from step 1 to $t-1$ are zero. In contrast, the turn angle θ depends only on the difference between the successive headings i.e. $\Theta_{t+1} - \Theta_t$. It is then possible to define a pair of ideal covariance functions.

The ideal allocentric covariance function is defined as

$$\begin{aligned} Cov_{allo} &= Cov(\Theta_t, \Theta_{t+1}) \\ &= \langle \Theta_t \Theta_{t+1} \rangle - \langle \Theta_t \rangle \langle \Theta_{t+1} \rangle . \end{aligned} \tag{2}$$

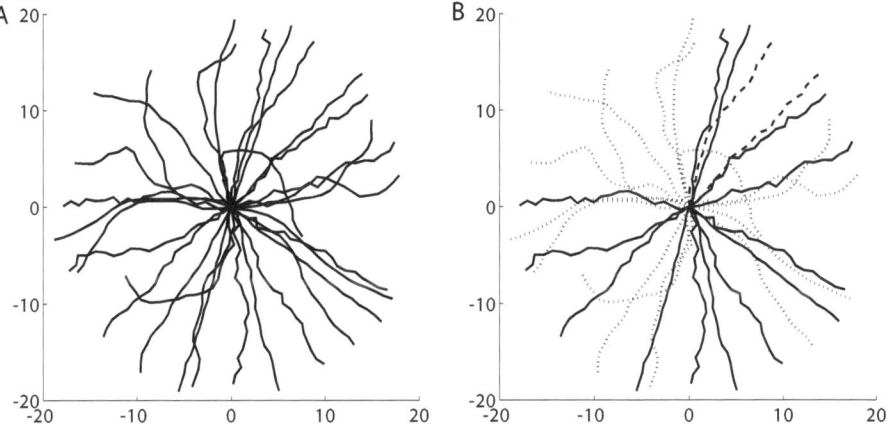

Fig. 1. (A) Graphical example of an unknown mixture of simulated (unbiased) idiothetic and allothetic directed walks with varying magnitudes of random angular displacement errors, and unknown axes of intended locomotion. (B) In this set of 30 paths of 20 steps, the decision function (Eqn 6) made the correct decision in 28 out of 30 paths (ADWs = solid lines, IDWs = dotted lines), being unable to decide in the remaining two (dashed lines). There were no incorrect decisions in this sample set.

Table 2. Angular components and covariance results of directed walks

Function	Type of Directed Walk IDW	ADW	Covariance Result $Cov\|IDW$	$Cov\|ADW$
Cov_{allo}	$\Theta_t = \sum_{j=1}^{t} \Delta_j$	$\Theta_t = \Delta_t$	$tV(\Delta)$	0
Cov_{ego}	$\theta_t = \Delta_t$	$\theta_t = \Delta_{t+1} - \Delta_t$	0	$-V(\Delta)$

The ideal egocentric covariance function is defined as

$$Cov_{ego} = Cov(\theta_t, \theta_{t+1})$$
$$= \langle \theta_t \theta_{t+1} \rangle - \langle \theta_t \rangle \langle \theta_{t+1} \rangle. \tag{3}$$

The pair of ideal covariance functions can be shown to have distinct angular error components and therefore different values when the directed walk is idiothetic or allothetic in nature. These results are summarized in Table 2. This implies that the pair of covariance values can be used to determine whether the directed walk was more likely to have been idiothetic or allothetic in nature.

3   Practical Application

In practice, systematic bias is removed by letting $\Theta'_t = \Theta_t - \overline{\Theta}$, which doesn't affect the turn angle $\theta'_t = \Theta'_{t+1} - \Theta'_t$. The following covariance estimates are used as inputs to the decision function (Eqn 6):

$$\widehat{Cov_{allo}} = \frac{1}{n-2} \sum_{t=1}^{n-1} (\Theta'_t - \overline{\Theta'}) (\Theta'_{t+1} - \overline{\Theta'}) \tag{4}$$

and

$$\widehat{Cov_{ego}} = \frac{1}{n-3} \sum_{t=1}^{n-2} (\theta'_t - \overline{\theta'}) (\theta'_{t+1} - \overline{\theta'}). \tag{5}$$

Careful examination reveals that the sample estimate of the allocentric covariance function in the case of IDWs is not an unbiased estimator (in contrast to the other three conditions). Nonetheless, it will be used for its practicality and simplicity. Using these values as inputs, it is then possible to decide whether the path travelled was more likely to be idiothetic (I), allothetic (A), or cannot be reliably decided (U). The decision function $\Lambda()$ is defined as follows:

$$\Lambda \left(\widehat{Cov_{allo}}, \widehat{Cov_{ego}} \right) = \begin{cases} A & if \left(\left| \frac{\widehat{Cov_{allo}}}{\widehat{Cov_{ego}}} \right| < 0.2 \right) \cap \left(\widehat{Cov_{ego}} < 0 \right) \\ I & if \left(\left| \frac{\widehat{Cov_{allo}}}{\widehat{Cov_{ego}}} \right| > 1 \right) \cup \left(\widehat{Cov_{ego}} > 0 \right) \\ U & Otherwise \end{cases} \tag{6}$$

Ideally, if the spatiotemporal resolution is sufficiently high to record each locomotory unit in detail, then it should be possible to distinguish each cycle of locomotion. It is then trivial to apply the decision function presented. However, if that is not practical, then it is better to undersample rather than oversample points along the journey. The main reason is to avoid the problem of a strong

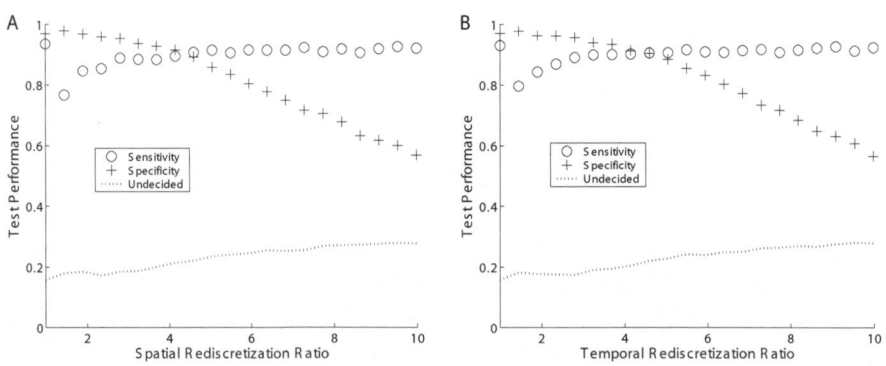

Fig. 2. Test performance of the decision function (Eqn 6) using simulated IDWs and ADWs. A range of spatial (A) and temporal (B) rediscretization ratios were used to downsample paths, keeping only the first 20 steps in each path. Simulation parameters: normally distributed Δ where σ_Δ was randomly chosen from the interval [0.1,0.5] rad; Δ_{bias} was also normally distributed with $\sigma_{bias} = 0.1$ rad. The performance results at each rediscretization ratio were calculated from 10,000 simulated paths. Note 'sensitivity' and 'specificity' were defined with respect to the ADW.

but spurious negative correlation between components within one locomotory unit. For example, if a leftward error is always followed by a rightward error within one elementary step, this would bias the statistics in favour of a decision of 'allothetic', irrespective of the directional cue used. In other words, the paths have to be rediscretized (Bovet and Benhamou 1988). In principle, this may be done either spatially or temporally as shown in Fig 2.

It can be seen from the simulation results that the method presented here can cope with a range of spatial (Fig 2A) and temporal (Fig 2B) rediscretization ratios, as well as tolerate noise and biases of varying magnitudes. It is noteworthy that performance (sensitivity and specificity) was always above chance with only 20 steps in each path sample, and yet the failure to make a decision rarely exceeded 20%.

4   Conclusions

1. *Population − based* statistical behaviours have already been described, and should be used where possible to determine the class of sensory input being used during animal navigation.

2. A novel method for quantifying *individual − based* statistical behaviour is under development and shows useful characteristics in simulations and early experimental trials (data not shown).

Acknowledgements. This work was supported by an NHMRC/ARC Thinking Systems grant and a COSNet Overseas Travel grant (COSNet Proposal number 93).

References

1. Bovet, P., Benhamou, S.: Spatial analysis of animals' movements using a correlated random walk model. J. Theor. Biol. 131, 419–433 (1988)
2. Cheung, A., Zhang, S., Stricker, C., Srinivasan, M.V.: Animal navigation: the difficulty of moving in a straight line. Biol. Cybern. 97, 47–61 (2007)
3. Pearson, K.: The problem of the random walk. Nature 72, 294 (1905)

Impact of Higher-Order Correlations on Coincidence Distributions of Massively Parallel Data

Sonja Grün[1,*], Moshe Abeles[2], and Markus Diesmann[1]

[1] Theoretical Neuroscience Group, RIKEN Brain Science Institute, 2-1 Hirosawa,
Wako-Shi, 351-0198 Saitama, Japan
[2] Gonda Brain Research Center, Bar Ilan University, Ramat Gan 52900, Israel
gruen@brain.riken.jp

Abstract. The signature of neuronal assemblies is the higher-order correlation structure of the spiking activity of the participating neurons. Due to the rapid progress in recording technology the massively parallel data required to search for such signatures are now becoming available. However, existing statistical analysis tools are severely limited by the combinatorial explosion in the number of spike patterns to be considered. Therefore, population measaures need to be constructed reducing the number of tests and the recording time required, potentially for the price of being able to answer only a restricted set of questions.

Here we investigate the population histogram of the time course of neuronal activity as the simplest example. The amplitude distribution of this histogram is called the complexity distribution. Independent of neuron identity it describes the probability to observe a particular number of synchronous spikes.

On the basis of two models we illustrate that in the presence of higher-order correlations already the complexity distribution exhibits characteristic deviations from expectation. The distribution reflects the presence of correlation of a given order in the data near the corresponding complexity. However, depending on the details of the model also the regime of low complexities may be perturbed.

In conclusion we propose that, for certain research questions, new statistical tools can overcome the problems caused by the combinatorial explosion in massively parallel recordings by evaluating features of the complexity distribution.

Keywords and Phrases: spike synchronization, higher-order synchrony, massively, parallel spike trains.

1 Introduction

Following the hypothesis that assembly activity is expressed by temporal relations between the spiking activity of the participating neurons, neuronal

---

* Corresponding author.

M. Marinaro, S. Scarpetta, and Y. Yamaguchi (Eds.): Dynamic Brain, LNCS 5286, pp. 96–114, 2008.
© Springer-Verlag Berlin Heidelberg 2008

responses need to be observed and analyzed with respect to temporal structure. With massively parallel recordings becoming available chances to observe the signature of assembly activity are increasing and indeed the availability of massively parallel spike data is escalating rapidly (e.g. Nicolelis et al., 1997; Csicsvari et al., 2003; Ikegaya et al., 2004). However, at this point in time we lack the corresponding analysis tools (Brown et al., 2004). Most of the existing methods are based on pairwise analysis (e.g. Aertsen et al., 1989; Nowak et al., 1995; Kohn & Smith, 2005; Shmiel et al., 2006), approaches to analyze correlations between more than two neurons do exist but typically work only for a small number of neurons (e.g. Abeles & Gerstein, 1988; Dayhoff & Gerstein, 1983; Grün et al., 2002a,b) or consider pair correlations only while analyzing the ensemble (e.g. Gerstein et al., 1985; Shlens et al., 2006; Schneidman et al., 2006). To extend existing methods designed to work on small number of neurons to massively parallel data is generally not feasible. One reason is that these methods typically assess individual spike patterns, e.g. coincidences with an identification of the participating neurons, or spatio-temporal spike pattern. An extension to many neurons would lead to a combinatorial explosion. This particularly holds for methods which include significance tests that do not only test against full independence, but detect higher-order correlations (e.g. Martignon et al., 1995; Nakahara & Amari, 2002; Schneider & Grün, 2003; Gütig et al., 2003; Ehm et al., 2007). Additional complications are the limited number of samples in experimental data, in particular if data are non-stationary. Only a few approaches exist that can handle and analyze massively parallel data for higher-order correlations. These approaches are based either on the model assumption of synfire chains (Schrader et al., 2008) or on compound Poisson processes (Staude et al., 2007, 2008).

Here we aim at a fast screening method that can detect correlation within massively parallel spike data. We base our approach on the distribution of the sum of spikes across neurons as reflected in the population histogram. In particular we explore how coincidence patterns of higher-order are reflected in this measure and if the order of the correlation can be identified. The dependence on parameters relevant for experimental data are studied using numerical and analytical methods.

Preliminary results have been presented in abstract form (Grün et al., 2003).

2 Correlation Model

We model massively parallel spike trains as N parallel stationary processes. For the generation of correlation we use slightly modified versions of two types of models, recently published by Kuhn et al. (2003). The basic idea underlying both models is to have a hidden Poisson 'mother' process of rate α, from which spikes are copied into parallel child processes according to a given probability ϵ. If $\epsilon = 1$ the model is named 'single interaction process' (SIP). All spikes of the mother process are present in all N child processes, such that synchronous higher-order spike events across all neurons are induced for each spike in the mother process.

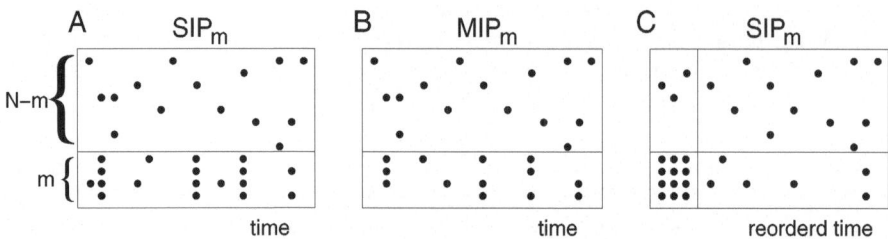

Fig. 1. Sketches of the models used for generating parallel spike data with higher-order spike coincidences in m out of N processes (A: SIP_m model, B MIP_m model). Rate p of the neurons is the same in both models and also for the m correlated and $N - m$ uncorrelated neurons. The bottom m neurons in the SIP_m model contain coincident events at rate α involving all m neurons. These neurons also contain uncorrelated background spikes at rate $p - \alpha$. In the MIP_m model the m neurons typically contain coincidences of lower order than m depending on the copy probability ϵ, and do not contain background spikes. The rate of insertion is $\alpha = p/\epsilon$. (C) shows the SIP_m model with reordered time bins. In the left part the bottom m neurons contain coincidences of order m in all time steps exhibiting the dependent fraction, the right part shows the independent fraction of the time bins.

In case $\epsilon < 1$ the model is called 'multiple interaction process' (MIP). Here, not necessarily all neurons receive a copy of the mother process' spikes, and the neurons receiving a spike are randomly chosen out of all N processes. As a result, the generated synchronous patterns differ in composition of neurons and in their complexity, i.e. the number of spikes in the pattern.

Here, we modify the models as follows. First, instead of expressing the model as Poisson processes in continuous time we formulate it in discretized time as Bernoulli processes. The reason is that we anyway aim to detect coincident events via binning. Formulating the model in continuous time with subsequent binning would lead to additional effects not relevant for the present study (cmp. Staude et al., 2008). Second, in a realistic recording session with N electrodes we do not expect to record from the same assembly at all electrodes. Instead we expect to observe at most a subset of $m < N$ neurons to participate in the same assembly. Therefore we copy the spikes of the mother process into m selected processes only. Third, correlated activity found in experimental data is typically embedded in uncorrelated 'background' firing activity (e.g. Riehle et al., 1997; Abeles et al., 1993; Prut et al., 1998). Therefore we 'dilute' the activity of the fully synchronized neurons with uncorrelated background spikes.

Thus, we define the following models of correlated activity.

Single interaction process in m out of N neurons: SIP_m. From a Bernoulli process of rate α, discretized in bins of width h (typically 1ms) we copy spikes with probability $\epsilon = 1$ into m out of N single neuron channels. Into each of the m processes of the correlated neurons we in addition to the synchronized events inject uncorrelated spikes modeled by a Bernoulli process of rate $p_b = p - \alpha$ (see Fig. 1A for a sketch of the model, and Fig. 2A for an example realization with

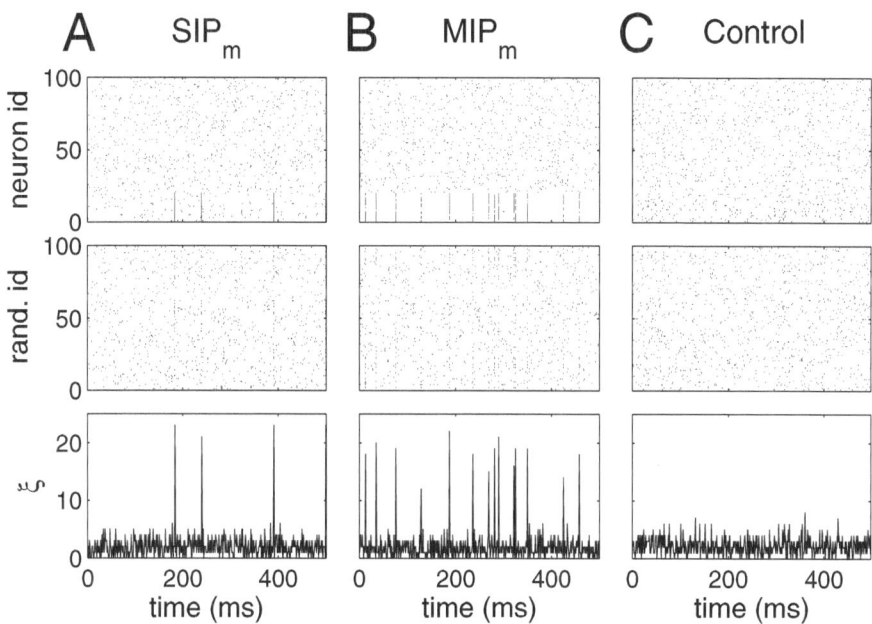

Fig. 2. Dot displays of realizations of the (A) SIP_m model, (B) MIP_m model and (C) control data generated by spike time randomization of the data of the MIP_m model in (B). The parameters common to all panels are $N = 100$, rate $p = 0.02$, and $h = 0.001\text{s}$. The duration shown is $T = 500ms$. In (A) neurons $1 \ldots 20$ ($m = 20$) are correlated with $\alpha = 0.005$, in (B) neurons $1 \ldots 20$ ($m = 20$) are correlated with $\epsilon = 0.8$. The middle row shows the same data as in the top row with random ordering of the neuron identifiers. The bottom row shows the population histograms (bin size $h = 0.001\text{s}$) identical for the top and the middle panel of each column.

ordered (top) and randomized (middle) neuron identifiers. Under the constraint that all N neurons have the same firing rate p we model the $N - m$ uncorrelated processes as independent Bernoulli processes of rate p. The parameter ranges we consider are the typical firing rates of cortical neurons (from a few to $\approx 100\text{Hz}$) and coincidence rates up to a few Hz as extracted from experimental cortical data (Grün et al., 1999). Thus α is typically small relative to the firing probability p.

Multiple interaction process in m out of N neurons: MIP_m. From a Bernoulli process of rate α, discretized in bins of width h (typically 1ms) we copy spikes with a probability $\epsilon \leq 1$ into m out of N neurons (for $\epsilon = 1$ this corresponds to SIP_m without background and $\alpha = p$). Here we do not insert background spikes into the m processes since due to the reduced copy probability also isolated spikes are generated appearing as background. The firing rate of these neurons is $p = \alpha \cdot \epsilon$. To fulfill the constraint that all neurons have the same firing rates the $N - m$ uncorrelated processes are modelled as a Bernoulli processes with firing probability p (see Fig. 1B for a sketch of the model, and

Fig. 2B for an example realization with ordered (top) and randomized (middle) neuron identifiers.

Control data. For comparison we generate control data without any correlated component. To this end we randomize the spike times (bins occupied by a spike) of SIP_m and MIP_m realizations along the temporal axis. This conserves the spike counts of the original data thereby avoiding additional variance (see Fig. 2C for an example of control data for the MIP_m model realization in Fig. 2B).

3   Distribution of Coincidence Counts

Next we explore how correlation between groups of neurons is expressed in simple measures like the population histogram and the distribution of the corresponding counts per bin. Thus we ignore individual spike constellations across the neurons and restrict the description to the distribution of the number of spikes of all N neurons within a bin, i.e. the complexity ξ (shown in Fig. 2, bottom row). We derive analytical descriptions of the distributions for both model types and then compare these to simulations.

3.1   Coincidence Count Distribution of SIP_m Model

The mathematical description of the coincidence distribution of correlated processes can be split into two contributions. Given that the time bins are independent, i.e. the processes have no memory, we can reorganize the bins in time (see Fig. 1C). Let us first collect the bins that contain injected coincidences ('dependent part'), then consider the rest i.e. all the time bins containing uncorrelated activity ('independent part'). The probability distribution of the dependent part is characterized by coincidences of order m with firing probability α in the m out of N neurons. The $N - m$ neurons fire independently with rate p. The coincidence distribution of only those $N - m$ neurons follows a binomial distribution

$$B(i, N - m, p) = \binom{N - m}{i} p^i \cdot (1 - p)^{(N - m - i)}. \tag{1}$$

These coincident events meet injected events of order m. Thus we obtain for the distribution of synchronized events of order $\xi = i + m$ in the N neurons:

$$P_{SIP,dep}(\xi) = B(\xi - m, N - m, p). \tag{2}$$

Since injected coincidences are of order m only patterns of complexities with at least m spikes are found; the probability for synchronous events of complexity $\xi < m$ is 0. The expectation value is

$$< \xi_{SIP,dep} > = (N - m) \cdot p + m. \tag{3}$$

The coincidence distribution of the remaining bins (independent part) is characterized by chance coincidences only. As before, the $N - m$ neurons exhibit chance coincidences according to a binomial distribution with firing probability p as expressed in Eq. 1. However, the firing probability of the m neurons in the independent part $p_b = p - \alpha$ is reduced by the injection probability of the synchronous events. The coincidence probability of these neurons is $B(j, m, p_b) = \binom{m}{j} p_b^j \cdot (1 - p_b)^{(m-j)}$. Consequently, we obtain the probability to observe a coincidence of a given complexity ξ by considering all possible combinations summing up to ξ of the number of coincidences among the group of $N - m$ neurons and the number of coincidences among the group of m neurons. This is expressed by the convolution of the two binomial distributions:

$$P_{SIP,indep}(\xi) = \sum_i B(i, N - m, p) \cdot B(\xi - i, m, p_b) \tag{4}$$

abbreviated as

$$= B(i, N - m, p) * B(\xi - i, m, p_b)$$

where we define $B(i, M, p) = 0$ for $i < 0$ and $i > M$. Thus the expectation value of this part of the complexity distribution is

$$< \xi_{SIP,indep} >= (N - m) \cdot p + m \cdot (p - \alpha) = Np - m\alpha. \tag{5}$$

Finally, the total coincidence distribution is the sum of the distributions of the two parts Eq. 2 and Eq. 4 weighted by the relative number of bins containing injected coincidences and its complement respectively. For large values of T we just consider the mean fractions α of the dependent part and $(1 - \alpha)$ of the independent part:

$$P_{SIP_m}(\xi) = \alpha \cdot P_{SIP,dep}(\xi) + (1 - \alpha) \cdot P_{SIP,indep}(\xi). \tag{6}$$

For simplicity we ignore the loss of spikes induced by the injection of coincident events into background activity with subsequent clipping (Grün et al., 1999) since for the range of parameters studied the probability for such collisions is very small $(\alpha \cdot p_b)$.

3.2 Coincidence Count Distribution of MIP$_m$ Model

In case of the MIP$_m$ model coincidences are inserted according to a given probability ϵ which defines the probability to copy spikes from the mother process into m of the N neurons. Let us again consider first only the part of the bins in which the mother process contained a spike (dependent part). The probability distribution for coincidences within the m neurons only is $B(i, m, \epsilon)$ (cmp. Kuhn et al., 2003). These events occur with the rate of the mother process, i.e. with probability α. The spiking probability of the m neurons is $p = \epsilon \cdot \alpha$. For the remaining $N - m$ neurons the same expression for the coincidence distribution holds as given above in Eq. 1. Thus, we yield for the dependent part:

$$P_{MIP,dep}(\xi) = B(i, m, \epsilon) * B(\xi - i, N - m, p). \tag{7}$$

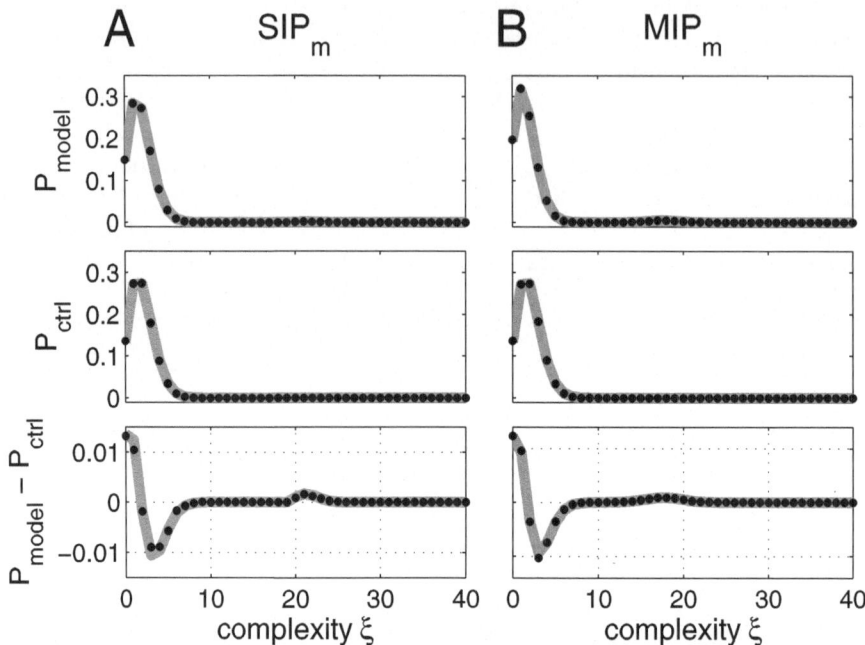

Fig. 3. Coincidence probability distributions of the SIP_m model (A) and the MIP_m model (B). Distributions (horizontal axis: coincidence complexity ξ ranging for better visibility only from 0 to 40) are constructed from realizations of duration $T = 100s$ at the resolution of the data $h = 0.001s$, with $N = 100$ neurons and firing probability $p = 0.02$ for all neurons. $m = 20$ neurons contain correlated activity. In case of SIP_m the coincidence rate is $\alpha = 0.005$, in case of MIP_m the copy probability is $\epsilon = 0.8$ resulting in an insertion probability of $\alpha = 0.025$. Top row: probability distribution of the raw coincidence counts; middle: results for the respective uncorrelated control data; bottom: difference distribution (model - control). The gray curves represent analytical results, black dots show results of simulations.

The mean complexity is

$$< \xi_{MIP,dep} >= (N - m) \cdot p + m \cdot \epsilon. \tag{8}$$

Thus, MIP_m differs from SIP_m in the sense that not all m neurons are correlated with order m. Furthermore, SIP_m contains background spikes in the correlated neurons, whereas MIP_m does not. For the part of the bins without a spike in the mother process (independent part) the description of the distribution is similar to the one for the SIP_m model (Eq. 4), however here the equation simplifies to chance coincidences from the $N - m$ neurons only:

$$P_{MIP,indep}(\xi) = B(\xi, N - m, p). \tag{9}$$

The mean complexity is

$$< \xi_{MIP,indep} >= (N - m) \cdot p. \tag{10}$$

The total probability distribution then results from weighting by the relative contributions:

$$P_{MIP,dep}(\xi) = \alpha \cdot P_{MIP,dep} + (1 - \alpha) \cdot P_{MIP,indep}$$
$$= \alpha \cdot B(i, m, \epsilon) * B(\xi - i, N - m, p) + (1 - \alpha) \cdot B(\xi, N - m, p).$$

Fig. 3 illustrates the coincidence count distributions for the SIP_m and the MIP_m model.

3.3   Control Data

In case of uncorrelated data (no insertion), the firing probability of all processes in Eq. 4 is p for both model types, and thus Eq. 4 reduces to:

$$P_{ctrl}(\xi) = B(i, m, p) * B(\xi - i, N - m, p) = B(\xi, N, p) \tag{11}$$

by using the addition formula for binomial coefficients. The same result holds for the control data for the MIP_m model. The expectation value for the mean complexity is $< \xi_{ctrl} >= Np$. This distribution reflects the coincidence distribution assuming full independence of the processes subject to the constraint of identical firing rates.

4   Comparing Model and Control Data

In the following we compare model and control data based on our analytical derivations and simulations. In particular we are interested to know in how far model and control data differ and, in view of data analysis, how this knowledge can be used to detect correlation. As the population dot displays in Fig. 2B demonstrate, model and control data are visually not distinguishable if the neuron identifiers are randomly arranged. Also the coincidence count distributions are visually very similar Fig. 3 (top, middle). Therefore we subtract the control data from the model data to highlight potential net excess coincidences in the model data Fig. 3 (bottom). Here deviations become clearly visible: at low complexities the model data contain more coincidences than the control, at slightly higher complexities they contain less coincidences, and for complexities at about $\xi = m$ there is again an excess of coincidences.

These features are characteristic for both models (Fig. 3, bottom row), however for SIP_m excess coincidences occur with a hump at a value of ξ slightly above m, for MIP_m the hump is located below $\xi = m$. Due to the chosen copy probability of $\epsilon = 0.8$ for MIP_m, the probability for patterns of complexity $\xi = m$ is low, and therefore the complexity of the hump is at $\xi < m$. In contrast, for SIP_m the hump is at values $\xi > m$, since patterns of complexity m are inserted, and by chance meet background spikes of the $N - m$ uncorrelated neurons.

The origin of the differences of the model data and the control data at low complexities is less obvious. Let us therefore restate the expression for the

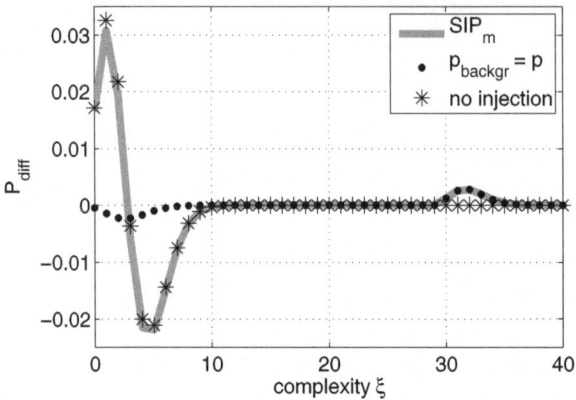

Fig. 4. Constituents of the complexity distribution. The thick gray curve shows the difference between the probability distribution of the SIP_m model and the respective control ($m = 30$, $N = 100$, $p = 0.03$, $\alpha = 0.01$, $T = 100s$, and $h = 0.001s$). The black dots show the difference distribution for a SIP_m model in which the background probability of the m neurons containing coincidences is elevated to match the rate p of the $N - m$ independent ones. Both SIP_m models exhibit the same excess hump starting at $\xi = 30$, but for the latter model the pronounced biphasic feature at low complexities is reduced to a small dimple (cmp. Eq. 4). The asterisks show the difference distribution for the original SIP_m model without injected coincidences; the m neurons only have uncorrelated background spikes at rate $p - \alpha$. The excess of high complexities is absent but the biphasic feature is conserved.

difference of the SIP_m model and the control data (subtract Eq. 3.3 from Eq. 6):

$$P_{SIP,diff}(\xi) = P_{SIP_m}(\xi) - P_{ctrl}(\xi)$$
$$= \alpha \cdot B(\xi - m, N - m, p) + (1 - \alpha) \cdot B(i, N - m, p) * B(\xi - i, m, p_b)$$
$$- B(\xi, N, p) \quad .$$

For the sake of simplicity we assume that coincidences are inserted into all N processes. Thus Eq. 4 simplifies to:

$$P_{SIP,diff}(\xi) = \alpha P_{dep} + (1 - \alpha)B(\xi, N, p_b) - B(\xi, N, p)$$

with

$$P_{dep} = \begin{cases} 0 & \text{if } \xi < N \\ 1 & \text{if } \xi = N \end{cases} \quad . \tag{12}$$

Injected coincidences enter $P_{SIP,diff}(\xi)$ only at $\xi = N$ with probability α. The remaining term expresses the contributions of the chance coincidences and their difference to the control data. Obviously, the distribution of the coincidences of the control data is shifted to a higher mean value (Np) as compared to the chance coincidences of the independent part of the model data (Np_b). The subtraction of the two distributions leaves a positive peak at small complexities, and a negative peak at somewhat higher complexities (Fig. 4). This difference can almost completely be compensated by increasing the rate of the independent part to p. In this case Eq. 4 reduces to:

$$P_{SIP,diff}(\xi) = \alpha P_{dep} + (1 - \alpha) \cdot B(\xi, N, p) - B(\xi, N, p)$$
$$= \alpha P_{dep} - \alpha \cdot B(\xi, N, p)$$

or in the general case of $m < N$

$$P_{SIP,diff}(\xi) = \alpha \cdot B(\xi - m, N - m, p) - \alpha \cdot B(\xi, N, p) \quad .$$

Whereas the entries at complexities $\xi > m$ are not affected Fig. 4 (cmp. the gray curve and black dots at the hump around $\xi = 32$), the biphasic feature at low complexities is reduced to a dimple reflecting the binomial distribution of chance coincidences scaled by $-\alpha$. The negative weight originates from the normalization constraint of the correlated data. The biphasic feature at low complexities can be replicated (Fig. 4, black asterisks) by independent data in which m neurons have a reduced firing probability $p - \alpha$ as it is the case in the independent part of the SIP_m model.

In conclusion, the biphasic feature at low complexities in the difference distribution of model and control data is due to the constraint of all neurons having the same firing probability and not due to the injected coincidences. The feature can be eliminated by adjusting the background rate of the correlated neurons to a rate comparable to the rate of the independent neurons. As outlined in the discussion section, we may be able to exploit this observation to differentiate between candidate mechanisms for the generation of correlated spiking in the neuronal system.

5   Parameter Variations

After having understood how inserted higher-order coincident spike events influence the coincidence distribution, we now study parameters relevant for the analysis of neurophysiological data. Typical questions on an experimental data set are: Is there correlation in the data? What is the order of the correlation, i.e. how many and which neurons are involved? Furthermore, synchronous spike events may occur with a temporal jitter that does not correspond to the bin width chosen for analysis.

5.1   Variation of Correlation Order m

In the SIP_m model coincidences of synchronized spike events of order m are inserted into m neurons and in Fig. 2A we studied the coincidence distributions

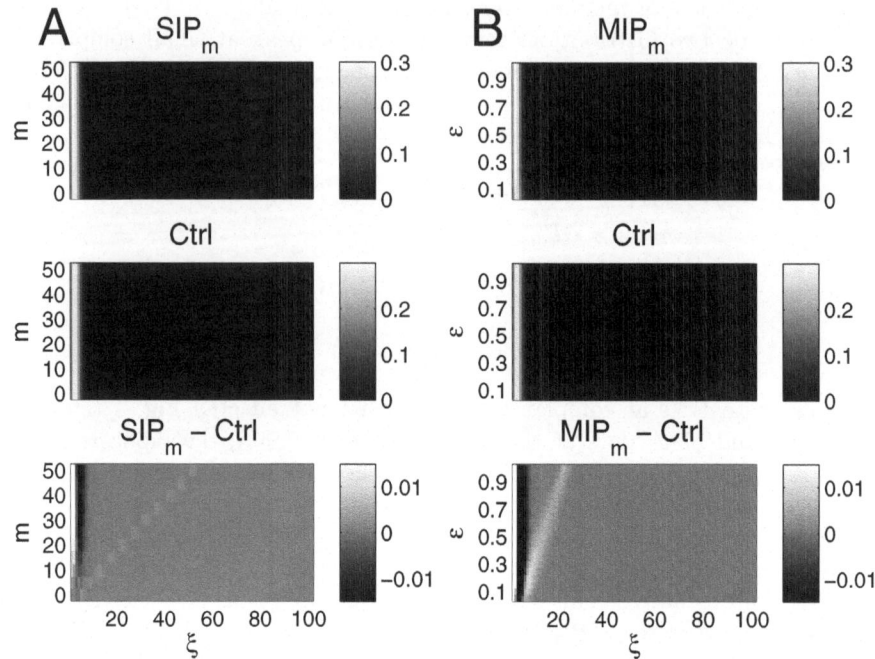

Fig. 5. Dependence of the complexity distribution on the order of the correlation (top: model data, middle: control data, bottom: difference of model and control). (A) variation (vertical) of m for the SIP_m model ($m = 5 \ldots 50$ in steps of 5, $\alpha = 0.005$) and (B) variation of ϵ for the MIP_m model ($\epsilon = 0.05 \ldots 1$ in steps of 0.05). The gray code indicates the probability to observe a coincidence pattern of a certain complexity ξ (horizontal). The data in the top panels result from simulations of the respective models with parameters $N = 100$, $p = 0.02$, $T = 100\text{s}$, and $h = 0.001\text{s}$. In the MIP_m model α is adjusted for each ϵ value to account for $p = \alpha \cdot \epsilon$. The control data are generated by temporal randomization of the spikes of the model data.

for $m = 20$ in $N = 100$ neurons. Now we are interested to see how the systematically varied order m of the injected coincidences affects the distribution. As before we study the distribution of the correlated data, the control data and their difference, however now visualized by a color code along the horizontal axis (ξ) and for increasing m along the vertical axis (Fig. 5A). Again, we find the biphasic feature for low complexities, which hardly varies with increasing m since the insertion and background rates are not changed.

The inserted coincidences, again, are not visible in the raw coincidence matrix. Only in the difference matrix (Fig. 5A,bottom) with increasing m excess coincidences appear at a complexity always somewhat higher than m since inserted coincidences of order m by chance meet background spikes which increases the complexity of the detected coincidence patterns.

For small m, the coincidences due to insertion overlap with the features due to the constraint on spike rate and the biphasic shape is disguised.

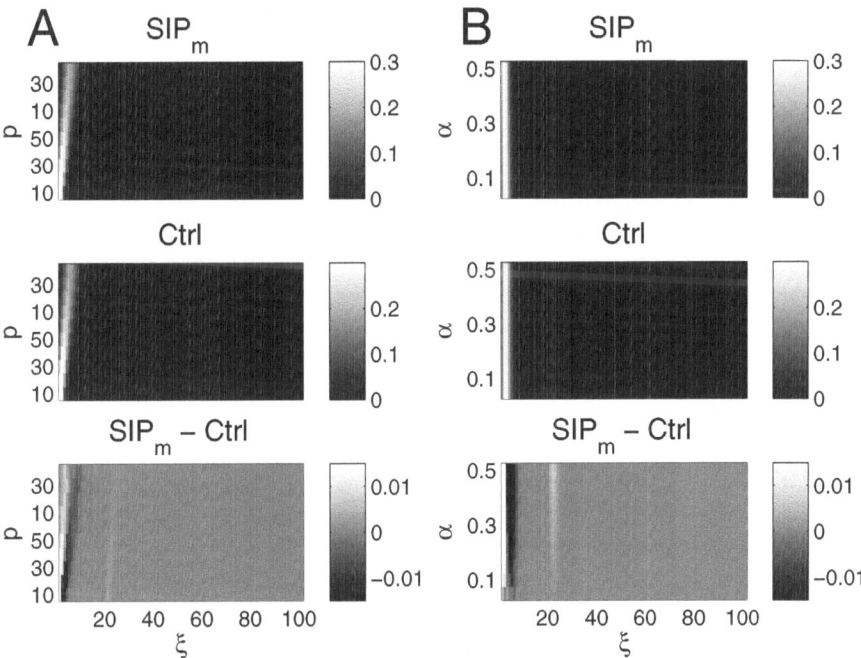

Fig. 6. Dependence of the complexity distribution on firing and coincidence probability in the SIP$_m$ model (top: model data, middle: control data, bottom: difference of model and control data). Common parameters of the simulations are: $m = 20$, $N = 100$, $T = 100$s, $h = 0.001$s. The gray code indicates the probability to observe a coincidence pattern of a certain complexity ξ (horizontal). (A) The firing probability $p = 0.005\ldots0.05$ is varied in steps of 0.05 with a constant coincidence probability of $\alpha = 0.05$. (B) The coincidence probability $\alpha = 0.002\ldots0.02$ is varied in steps of 0.002 at a constant firing probability of $p = 0.02$.

5.2 Variation of Copy Probability ϵ

In physiological terms the copy probability ϵ of the MIP$_m$ model corresponds to the participation probability of the neurons in an assembly activation. In Fig. 5B we keep the total number of neurons N constant, as well as the number of neurons m (here 20) in which coincidences are inserted. With increasing ϵ the mean complexity of the excess coincidences increase linearly according to Eq. 8. For $\epsilon = 1$, which corresponds to SIP$_m$ without background, the mean complexity reaches a value slightly above m (Fig. 5B, bottom). The amplitude of the hump decreases with increasing ϵ which is due the requirement of constant firing rate subject to the relation $p = \epsilon \cdot \alpha$. To fulfill this constraint the rate of the mother process α has to decrease with increasing ϵ. At the same time the variance of the complexity of the excess coincidences becomes larger according to $m \cdot \epsilon \cdot (1 - \epsilon)$), but due to the simultaneous decrease of α the hump appears less wide in the difference plot.

Summarizing the variation of m and ϵ leaves us with the insight that from the position of the excess hump we cannot directly conclude on the number of neurons involved in the correlation if the underlying model is not known. The hump position could either directly reflect the number of correlated neurons for SIP_m, or is indicating a smaller number of neurons than are actually involved in the correlation due a value of $\epsilon < 1$ in the MIP_m model. As a consequence the underlying model must first be identified. One option is to extract the coincidence patterns which exhibit excess complexities and analyze them for their individual composition. If always the same set of neurons is active in a pattern we can conclude on SIP_m. However, if patterns are composed of subsets of a particular superset of neurons we can conclude on a MIP_m type model.

5.3   Variation of Firing Rates

Fig. 6A, top panel shows the dependence of the coincidence distribution on the total firing probability p for constant N, m, and coincidence probability α in case of SIP_m. By increasing p also the background spike probability p_b of the neurons with injected coincidences increases ($p_b = p - \alpha$). Consequently, the mean complexity of the peak of the independent part of the coincidence distribution increases according to Eq. 5. The complexity at the hump of the excess coincidences (Fig. 6A, bottom) mainly corresponding to the dependent part also increases with the firing probability p but with a smaller slope, i.e. according to Eq. 3 with $(N - m)p + m$, since $(N - m)p < Np$.

Similar considerations hold for changes of the coincidence rate α (Fig. 6B). Keeping all other parameters constant, a change of the coincidence injection probability α only affects the hump height, but not its complexity. Only the complexity of the independent part at small complexities is affected, since an increase of the coincidence probability leads to a decrease of the background probability in the m neurons. Consequently, as discussed in section 4, the difference in chance coincidences of the model and the control data increases (cmp. Eq. 4).

5.4   Variation of Bin Width vs. Temporal Jitter

Coincident spike events of pairs of cortical neurons typically have a temporal jitter of a few ms (see e.g. Grün et al., 1999; Pazienti et al., 2008). Such a jitter can be modeled by copying the spikes of the mother processes not always into exactly the same bin across the neurons, but to allow copying into neighboring bins with some probability. For the sake of simplicity, here we decide for a rectangular distribution of spike times as described in the caption of Fig. 7.

One option to detect jittered coincidences, is to adjust the bin width (e.g. Grün et al., 1999, 2002a) to the precision of the spikes. Since in an electrophysiological experiment the appropriate bin width cannot be known in advance, we analyze the same data set with increasing bin width. Thus, instead of counting the simultaneously emitted spikes at the resolution of the data h, we now count the spikes within w neighboring time steps. w is called the bin width and we

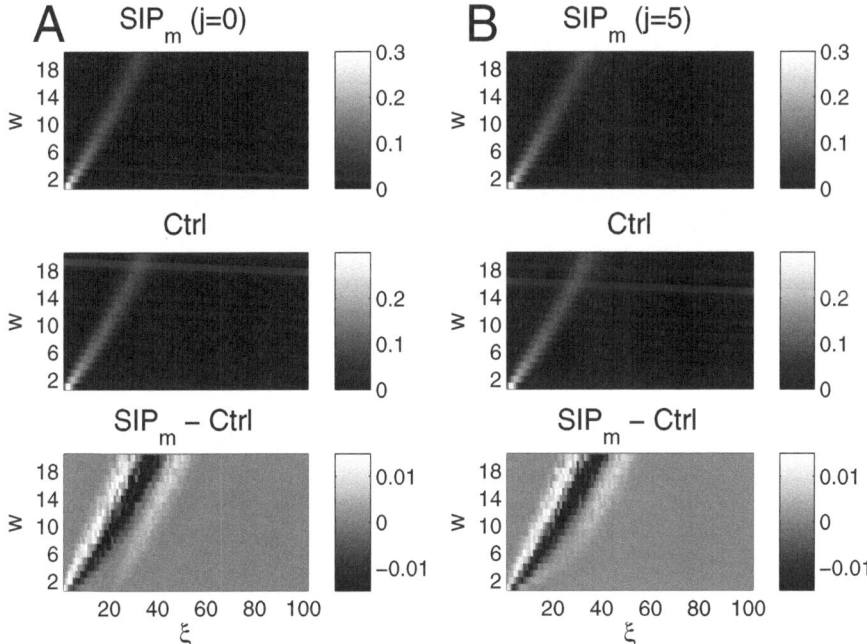

Fig. 7. Coincidence probability distribution (top: model data, middle: control data, bottom: difference of model and control) of the SIP_m model under variation of the bin width w (varied from 1 to 20 in steps of 1 in units of $h = 0.001$s, vertical axis). The analyzed data sets differ in the temporal jitter of the inserted coincidences: in (A) the coincidences are exact without temporal jitter, in (B) the jitter is $j = 5h$. The jitter is generated by randomly displacing each spike of each neuron within a window of $\pm 0.5j$ centered at its original position, thereby generating coincident spike events with a maximal distance of j. The gray code indicates the probability of occurrence of coincidence patterns as a function of complexity ξ (horizontal). The data in the top panels result from simulations of the SIP_m model with parameters $N = 100$, $T = 100$s, $h = 0.001$s, $m = 20$, $\alpha = 0.05$ and $p = 0.02$.

also call the sum of events complexity although a single neuron may contribute more than one spike.

Similar to the foregoing displays we then plot the coincidence probability distribution as a function of the complexity ξ horizontally, and along the vertical axis as a function of the bin width w (in units of h). Fig. 7 shows in A the result for a non-jittered data set and in B the result for a data set where coincidences have a jitter of up to $j = 5$ms. Even in the case of no jitter the complexity of the peak of the chance coincidences increases with bin width. This holds for both, the model and the control data. The reason is that with increasing bin width the probability to detect spikes within a bin trivially increases due to background activity. For the independent part the coincidence probability distribution for a bin width w reads:

$$P_{SIP,indep}(\xi, w) = (1 - \alpha) \cdot B(i, (N - m) \cdot w, p) * B(\xi - i, m \cdot w, p_b)$$

Thus the mean of the independent part of the complexity distribution increases as $w(N - m) \cdot p + wm \cdot p_b)$. The detected complexities are generally higher than in the original case where the width of the bins corresponds to the resolution h of the generating Bernoulli process.

For the dependent part we obtain:

$$P_{SIP,dep}(\xi, w) = \alpha \cdot B(\xi - m, (N - m)w, p). \tag{13}$$

The mean complexity is $m + w(N - m)p$.

If coincidences are jittered, only part of the coincidences are detected for $wh < j$. For small w the complexity of the detected excess coincidences is reduced to the number of spikes of an injected coincidence falling into one bin. Furthermore, the counts of these events is smaller than the injected number and therefore the hump of the excess coincidences is small. From $wh = j$ on, all inserted coincidences are detected (neglecting debris due to exclusive binning; Grün et al., 1999). The complexity of the hump is now increasing faster with w as for the range of bin sizes for $w < j$. This change of slope at $w = j$ may be used as an indicator for the precision of synchronization in the data.

6 Discussion

The goal of this study is to learn if a simple measure like the population histogram generated from parallel spike data can be used to detect correlation between the spike trains. To study such a situation we re-formulated two types of models for parallel point processes, which enables us to generate many parallel spike trains with a subgroup of them being correlated and exhibiting coincident spiking. As a measure of correlation between the spike trains we use the coincidence count distribution which is the amplitude distribution of the population histogram, i.e. the sum of spikes across the neurons within a predefined bin size as a function of time.

We illustrate that correlations are neither directly visible in the population dot display nor in the coincidence distribution, since background spikes act as a strong noise component. These spikes generate a large number of entries in the coincident count distribution at low complexities dominating the distribution. We choose to account for effects of background firing rate by comparing the model data to control data, which contain the same number of spikes per neuron but with the single neurons exhibiting fully independent firing. Still, the distribution of the original and the control data appear very similar by visual inspection, and only the difference of the two gives indication of correlation in the model data.

We find that correlation in a subset of the neurons can be clearly identified by entries in the difference matrix at the complexities close to the order of correlation contained in the data. The exact complexity observed depends on the assumed underlying model. For SIP_m the minimum complexity of the excess coincidence entries reflects the number of correlated neurons, for MIP_m the maximum. The two models can be distinguished by extracting the time bins of entries of high complexity and analyzing the detailed composition of the neurons contributing to the patterns. In SIP_m a subset of m neurons would always fire together, for MIP_m a subset of m neurons would be involved but would contribute with different compositions of neurons. The latter seems to be a quite realistic assumption. Realizations of the synfire chain model would exhibit such activity in each group of the chain, since not all neurons in a group need to be active for stable propagation of a synfire run (Diesmann et al., 1999). A typical participation probability for each individual neuron, which corresponds to ϵ in the MIP_m model, is about 0.8.

If the number of correlated neurons, however, becomes small compared to the chance complexities the hump of the excess coincidences drowns in the noise. However, for m as small as 5 the hump of the excess coincidences already separates from the background. Larger background firing rate would indeed shift the peak of the chance coincidences to somewhat higher complexities, but also shifts the hump of the excess coincidences due to enhanced chances to meet background spikes.

An obvious biphasic feature in the difference coincidence distribution occurs at low complexities which we identified as being due to the constraint of all neurons having the same firing probability. As a consequence of the presence of correlated spikes the background firing of the neurons in the correlated data set is reduced. Therefore the distribution exhibits a decreased probability for small complexities as compared to the fully independent control data. Thus, if coincident events are injected into processes containing the same rate as the independent processes, the biphasic feature is absent and the difference coincidence distribution becomes almost flat in this regime.

This observation provides an interesting option to distinguish two potential mechanisms for the generation of synchronous spike events in the neuronal system. In one scenario additional synchronized events are generated by the system without affecting the ongoing activity, which would correspond to the case where the neurons receiving additional coincidences have an increased rate (by $p-p_b$) as compared to the neurons without coincidences. An example situation is where some of the observed neurons are part of an occasionally active feed-forward subnetwork (synfire chain, Schrader et al., 2008). Alternatively, spikes could be shifted such that they become synchronized with others. This would correspond to the case where firing rates are the same for all neurons. An example situation is where spikes become locked to global network oscillations.

The comparison of experimental data with data that implement control models is a standard approach in the correlation analysis of parallel spike trains (e.g. Pipa & Grün, 2003; Ikegaya et al., 2004; Shmiel et al., 2006). Such a comparison is mostly formulated as a significance test where the control data realize a specific null-hypothesis. The analysis presented here is rather thought of as a fast scanning procedure helping to decide if a data set contains interesting correlation which should further be analyzed in more detail. It provides a comparison of original data to control data realizing full statistical independence. The latter can be analytically described by a Bernoulli distribution. However, also a numerical realization can be achieved easily in practice by randomizing the bins of each neuron in time. Such a surrogate also conserves the spike count of each neuron. In contrast, realizations of Bernoulli processes with the firing probability estimated from the data would introduce additional variance.

A next step would be to add a statistical test to the discussed approach. One way of quantifying differences of the complexity distributions would be to perform a test on the full distribution (e.g. Pipa & Grün, 2003; Pipa et al., 2008) another to calculate and compare the moments or cumulants (Staude et al., 2007, 2008). The latter also provides the option to compare the experimental data to models which include sucessively higher orders of correlation as implemented in Staude et al. (2008), which then enable statements on the minimal order consistent with the data.

Other aspects that need to be considered in extentions of the current approach are typical features of experimental data: non-stationarity of the firing rate in time, inhomogeneous rates of the different neurons, and temporal modulation of the correlation (e.g. Riehle et al., 1997). Such situations can rapidly reach a level of complexity severely impeding analytical descriptions but the flourishing idea of surrogates can come to the rescue (see Grün (2008) for a review).

Acknowledgments. This work was initiated during a scientific stay of SG and MD in 2002 at the Hadassah Medial School of the Hebrew University in Jerusalem financed by the NEURALCOMP-ICNC program. The project was supported in part by the Volkswagen Foundation, the Stifterverband für die deutsche Wissenschaft, the BCCN Berlin and Freiburg (BMBF grants 01GQ01413 and 01GQ0420), and DIP F1.2.

References

Abeles, M., Bergman, H., Margalit, E., Vaadia, E.: Spatiotemporal firing patterns in the frontal cortex of behaving monkeys. J. Neurophysiol. 70(4), 1629–1638 (1993)

Abeles, M., Gerstein, G.L.: Detecting spatiotemporal firing patterns among simultaneously recorded single neurons. J. Neurophysiol. 60(3), 909–924 (1988)

Aertsen, A.M.H.J., Gerstein, G.L., Habib, M.K., Palm, G.: Dynamics of neuronal firing correlation: Modulation of 'effective connectivity'. J. Neurophysiol. 61(5), 900–917 (1989)

Brown, E.N., Kaas, R.E., Mitra, P.P.: Multiple neural spike train data analysis: state-of-the-art and future challenges. Nat. Neurosci. 7(5), 456–461 (2004)

Csicsvari, J., Henze, D.A., Jamieson, B., Harris, K.D., Sirota, A., Barth, P., Wise, K.D., Buzsaki, G.: Massively parallel recording of unit and local field potentials with silicon-based electrodes. J. Neurophysiol. 90, 1314–1323 (2003)

Dayhoff, J.E., Gerstein, G.L.: Favored patterns in spike trains. I. detection. J. Neurophysiol. 49(6), 1334–1348 (1983)

Diesmann, M., Gewaltig, M.-O., Aertsen, A.: Stable propagation of synchronous spiking in cortical neural networks. Nature 402(6761), 529–533 (1999)

Ehm, W., Staude, B., Rotter, S.: Decomposition of neuronal assembly activity via empirical de-poissonization. Electron. J. Statist. 1, 473–495 (2007)

Gerstein, G.L., Perkel, D.H., Dayhoff, J.E.: Cooperative firing activity in simultaneously recorded populations of neurons: Detection and measurement. J. Neurosci. 5(4), 881–889 (1985)

Grün, S.: Data driven significance estimation for precise spike correlation (invited review). J. Neurophysiol. (submitted 2008)

Grün, S., Abeles, M., Diesmann, M.: The impact of higher-order correlations on coincidence distributions of massively parallel data. In: Proc. 5th Meeting German Neuroscience Society, pp. 650–651 (2003)

Grün, S., Diesmann, M., Aertsen, A.: 'Unitary Events' in multiple single-neuron spiking activity. I. Detection and significance. Neural Comput. 14(1), 43–80 (2002a)

Grün, S., Diesmann, M., Aertsen, A.: Unitary Events in multiple single-neuron spiking activity. II. Non-Stationary data. Neural Comput. 14(1), 81–119 (2002b)

Grün, S., Diesmann, M., Grammont, F., Riehle, A., Aertsen, A.: Detecting unitary events without discretization of time. J. Neurosci. Methods 94(1), 67–79 (1999)

Gütig, R., Aertsen, A., Rotter, S.: Analysis of higher-order neuronal interactions based on conditional inference. Biol. Cybern. 88(5), 352–359 (2003)

Ikegaya, Y., Aaron, G., Cossart, R., Aronov, D., Lampl, I., Ferster, D., Yuste, R.: Synfire chains and cortical songs: temporal modules of cortical activity. Science 5670(304), 559–564 (2004)

Kohn, A., Smith, M.A.: Stimulus dependence of neuronal correlations in primary visual cortex of the Macaque. J. Neurosci. 25(14), 3661–3673 (2005)

Kuhn, A., Aertsen, A., Rotter, S.: Higher-order statistics of input ensembles and the response of simple model neurons. Neural Comput. 1(15), 67–101 (2003)

Martignon, L., von Hasseln, H., Grün, S., Aertsen, A., Palm, G.: Detecting higher-order interactions among the spiking events in a group of neurons. Biol. Cybern. 73, 69–81 (1995)

Nakahara, H., Amari, S.: Information-geometric measure for neural spikes. Neural Comput. 14, 2269–2316 (2002)

Nicolelis, M., Ghazanfar, A., Faggin, B., Votaw, S., Oliverira, L.: Reconstructing the engram: simultaneous, multisite, many single neuron recordings. Neuron 18(4), 529–537 (1997)

Nowak, L.G., Munk, M.H., Nelson, J.I., James, A., Bullier, J.: Structural basis of cortical synchronization. I. Three types of interhemispheric coupling. J. Neurophysiol. 74(6), 2379–2400 (1995)

Pazienti, A., Diesmann, M., Grün, S.: The effectiveness of systematic spike dithering depends on the precision of cortical synchronization. Brain Research 1225, 39–46 (2008)

Pipa, G., Grün, S.: Non-parametric significance estimation of joint-spike events by shuffling and resampling. Neurocomputing 52–54, 31–37 (2003)

Pipa, G., Wheeler, D., Singer, W., Nikolic, D.: Neuroxidence: Reliable and efficient analysis of an excess or deficiency of joint-spike events. J. Comput. Neurosci. 25(1), 64–88 (2008)

Prut, Y., Vaadia, E., Bergman, H., Haalman, I., Hamutal, S., Abeles, M.: Spatiotemporal structure of cortical activity: Properties and behavioral relevance. J. Neurophysiol. 79(6), 2857–2874 (1998)

Riehle, A., Grün, S., Diesmann, M., Aertsen, A.: Spike synchronization and rate modulation differentially involved in motor cortical function. Science 278(5345), 1950–1953 (1997)

Schneider, G., Grün, S.: Analysis of higher-order correlations in multiple parallel processes. Neurocomputing 52–54, 771–777 (2003)

Schneidman, E., Berry, M.J., Segev, R., Bialek, W.: Weak pairwise correlations imply strongly correlated network states in a neural population. Nature 440, 1007–1012 (2006)

Schrader, S., Grün, S., Diesmann, M., Gerstein, G.: Detecting synfire chain activity using massively parallel spike train recording (in press, 2008)

Shlens, J., Field, G.D., Gauthier, J.L., Matthew, I.P.D., Sher, A., Litke, A.M., Chichilnisky, E.: The structure of multi-neuron firing patterns in primate retina. J. Neurosci. 26(32), 8254–8266 (2006)

Shmiel, T., Drori, R., Shmiel, O., Ben-Shaul, Y., Nadasdy, Z., Shemesh, M., Teicher, M., Abeles, M.: Temporally precise cortical firing patterns are associated with distinct action segments. J. Neurophysiol. 96(5), 2645–2652 (2006)

Staude, B., Rotter, S., Grün, S.: Detecting the existence of higher-order correlations in multiple single-unit spike trains. In: Society for Neuroscience, Volume 103.9/AAA18 of Abstract Viewer/Itinerary Planner, Washington, DC (2007)

Staude, B., Rotter, S., Grün, S.: Inferring assembly-activity from population spike trains (submitted, 2008)

Comparing Kurtosis Score to Traditional Statistical Metrics for Characterizing the Structure in Neural Ensemble Activity

Peter Stratton[1,2] and Janet Wiles[1,2]

[1] School of Information Technology and Electrical Engineering
[2] Queensland Brain Institute,
University of Queensland, Australia
{stratton,wiles}@itee.uq.edu.au
http://www.itee.uq.edu.au/~stratton/

Abstract. This study investigates the range of behaviors possible in ensembles of spiking neurons and the effect of their connectivity on ensemble dynamics utilizing a novel application of statistical measures and visualization techniques. One thousand spiking neurons were simulated, systematically varying the strength of excitation and inhibition, and the traditional measures of spike distributions – spike count, ISI-CV, and Fano factor – were compared. We also measured the kurtosis of the spike count distributions. Visualizations of these measures across the parameter spaces show a range of dynamic regimes, from simple uncorrelated spike trains (low connectivity) through intermediate levels of structure through to seizure-like activity. Like absolute spike counts, both ISI-CV and Fano factor were maximized for different types of seizure states. By contrast, kurtosis was maximized for intermediate regions, which from inspection of the spike raster plots exhibit nested oscillations and fine temporal dynamics. Brain regions exhibit nested oscillations during tasks that involve active attending, sensory processing and memory retrieval. We therefore propose that kurtosis is a useful addition to the statistical toolbox for identifying interesting structure in neuron ensemble activity.

1 Introduction

With the multi-electrode recording techniques available today, it is possible to discern spiking activity from dozens or even hundreds of neurons simultaneously from awake, behaving animals [1]. In the first half of the twentieth century, it was discovered that electrical activity in the brain oscillates in characteristic ways (see [2] for a review at the time), and now these new recording techniques have allowed the examination of neural ensemble spiking activity, and how it relates to local field potential (LFP) recordings which characterise brain oscillations. It can be seen that simulated populations of neurons also display synchronous and oscillatory behaviour (see Figure 1), and theoretical work has shown how this behaviour can be supported by the individual firing regimes of sparsely connected neurons [3].

M. Marinaro, S. Scarpetta, and Y. Yamaguchi (Eds.): Dynamic Brain, LNCS 5286, pp. 115–122, 2008.
© Springer-Verlag Berlin Heidelberg 2008

There are numerous methods for measuring regularities in spike trains (e.g. power spectra, spike-count distributions). Two of the most popular are the Fano Factor [4] and the coefficient of variation of the interspike interval (ISI-CV) [5]. These two measures are maximised for super-synchronous seizure-like spike trains, and in simulations of coupled networks of neurons, are themselves closely correlated with the total number of spikes generated by the network in a given time. Hence, they do not provide a measure of the structure in realistic spike trains with typical intermediate levels of activity. The new metric presented here, called the *Kurtosis Score*, more clearly reveals the nested oscillatory patterns of activation found in simulated spike trains, where bursts of high frequency appear at the peaks of lower frequency oscillations.

With desktop computing power available today, it is becoming practical to use *parameter sweeps* across multi-dimensional spaces as a tool for investigating the behaviour of complex systems. If system behaviours can be quantified with one or more metrics, these metrics can be displayed as graphs called *heat maps* that can reveal relationships between parameters and between metrics that may otherwise remain hidden. In the following sections, heat maps are used to compare Kurtosis Score, total spike count, Fano Factor and ISI-CV for their effectiveness at identifying simulated spike trains with varying characteristics. The chapter concludes with a discussion of when it may be best to use Fano Factor or ISI-CV, and when Kurtosis Score may be the more appropriate metric.

2 Methods

Kurtosis Score is defined as the fourth cumulant divided by the square of the variance of the spikes per millisecond distribution, simultaneously recorded or simulated for a large number of neurons. Fano Factor [4] is the variance divided by the mean of the spikes per millisecond distribution. ISI-CV [5] is the standard deviation divided by the mean of the interspike interval (interspike intervals are calculated for each neuron separately, and then all calculated intervals are combined for the ISI distribution).

In order to compare these metrics, Kurtosis Score, spike count, Fano Factor and ISI-CV were calculated for simulated spike train data. This data was obtained by simulating networks of Izhikevich model neurons [6]. Each network was fully connected, contained 800 pyramidal cells and 200 interneurons and was run for 1000 ms of simulated time. The metric calculations excluded the first 400 ms of each trial to give the networks time to settle from their initial conditions; the metrics were than calculated for the time period 401 to 1000 ms. The excitatory weights in the networks were distributed (uniform random) between zero and a maximum weight w^+_{max} that varied from trial to trial. w^+_{max} varied systematically between zero and a value large enough for approximately four presynaptic neuron spikes to alone cause a spike in a postsynaptic neuron (typical cortical synaptic efficacies require 10 to 40 presynaptic spikes to activate a postsynaptic neuron, which is approximately in the middle of the range over which w^+_{max} varied). The maximum inhibitory synaptic weight w^-_{max} varied between zero and twice the maximum excitatory weight (i.e. $w^-_{max} = -2w^+_{max}$). All weights were subject to short term depression with exponential recovery [7]. The excitatory and inhibitory weight variation between trials created networks that behaved very differently across different regions of this parameter space, and the spike trains thus generated ranged from uncorrelated activity to super-synchronous seizure-like states. Each neuron received zero-mean Gaussian input, with standard deviation of 5 for pyramidal cells and 2 for interneurons [6]. Because the network input and connections are stochastic, 50 to 100 instances of the networks were simulated at each testing point in parameter space and the calculated metrics averaged to obtain a picture of the mean network behaviour at each parameter space point. The averaged metric scores were plotted on 2-dimensional graphs called heat maps for visualisation of how the metrics change over the parameter space, with excitatory weight on the abscissa and inhibitory weight the ordinate.

3 Results

Simulations of the network with varying excitatory and inhibitory connection strengths showed a variety of behaviours (see Figure 1). In these examples, Fano Factor and ISI-CV were maximal for super-synchronised spike trains, while Kurtosis Score was maximal for spike trains containing bursts of gamma wave activity nested within slower oscillations. By systematically varying the parameters for the excitatory and inhibitory connection strengths, a picture of how these metrics change and relate to each other can be constructed (see Figure 2; full color version available online). When Kurtosis Score was mapped across the entire tested parameter space, a region of high kurtosis where spike trains exhibit high structure became apparent (Figure 2D – light blue to white regions). When compared with ISI-CV, Fano Factor and the total number of spikes generated in the simulation period, Kurtosis Score can be seen to be high in regions of parameter space adjacent to high ISI-CV and Fano Factor, where the total number of spikes also begins rising dramatically (Figure 2A, B and C). Clearly high Fano Factor and ISI-CV primarily indicate super-synchrony (which also tends to generate more total spikes), while Kurtosis Score is quite low in these super-synchronous regions of weight space. Instead, Kurtosis Score is high when the spike train contains fine but unpredictable temporal structure, rather than either fully predictable or completely random.

4 Discussion

Fano Factor and ISI-CV are both ratio metrics with the mean as the divisor. Thus these values are maximized as the mean approaches zero, as long as some variance is maintained. For Fano Factor this maximization occurs in a seizure-like state where bursts of activity are short and 'rest' periods long, while for ISI-CV it occurs also in seizure but when bursts are long and rest times short. If the Fano Factor and ISI-CV graphs from Figure 2 are rescaled so that the shade transitions occur in the region where Kurtosis Score is highest, it can be seen that ISI-CV actually decreases, while Fano Factor increases monotonically towards the seizure state throughout the region of high kurtosis (see Figure 3). The clear implication is that Fano Factor and ISI-CV are not good measures of spike train structure; rather they give slightly different accounts of the closeness of a spike train to super-synchrony.

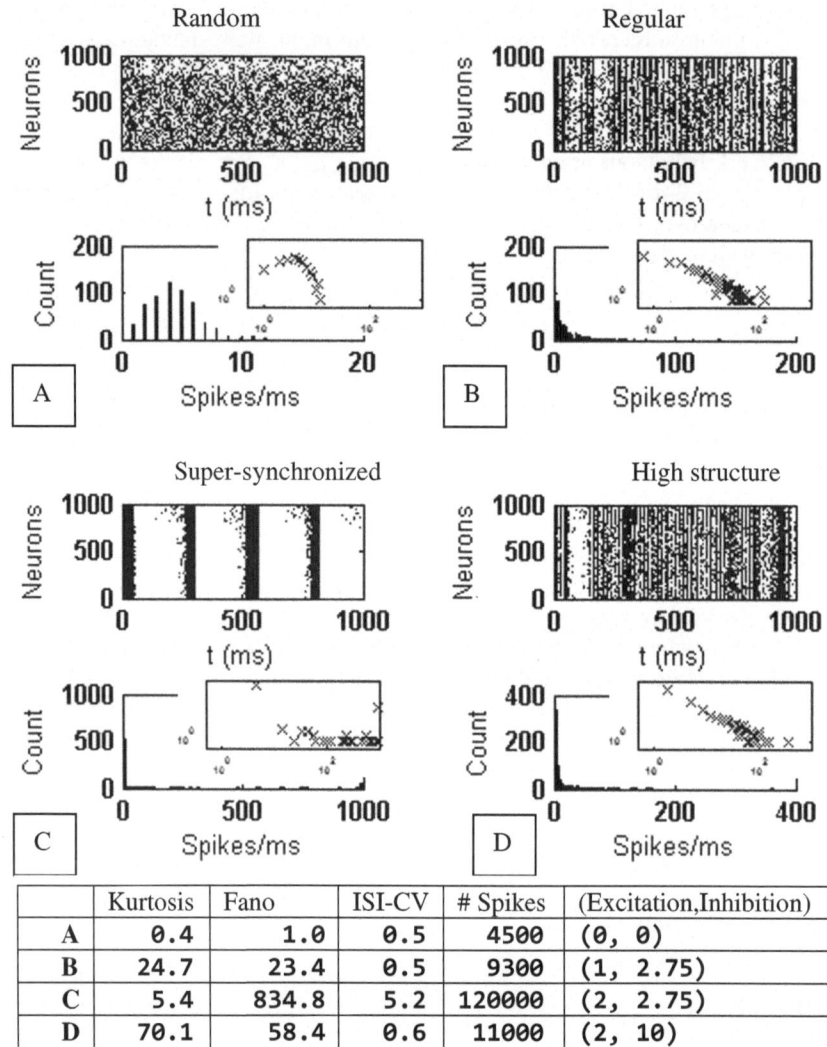

Fig. 1. Characteristic behaviors from different regions of parameter space. Spike raster plots from four simulated networks show the same 100 neurons fully connected with different synaptic strengths, resulting in very different network dynamics in each case. Below each raster plot is the spike count distribution. *(A)*: Random. No connections between neurons gives a Kurtosis Score close to 0, a Fano Factor close to 1, indicating a near Poisson distribution, and a low ISI-CV. *(B)*: Regular. Intermediate excitation and inhibition gives a moderately high Kurtosis Score, a higher but still relatively small Fano Factor and a low ISI-CV. *(C)*: Super-synchronized. Strong excitation and intermediate inhibition gives a low Kurtosis Score but Fano Factor and ISI-CV are maximized by this super-synchronized activity. *(D)*: High structure. Strong excitation and strong inhibition strengthens the fast oscillations and 'fine' temporal dynamics, and hence maximizes the Kurtosis Score, while Fano Factor and ISI-CV are reduced. *(Insets)*: The spike count distributions plotted on log-log axes (see Discussion).

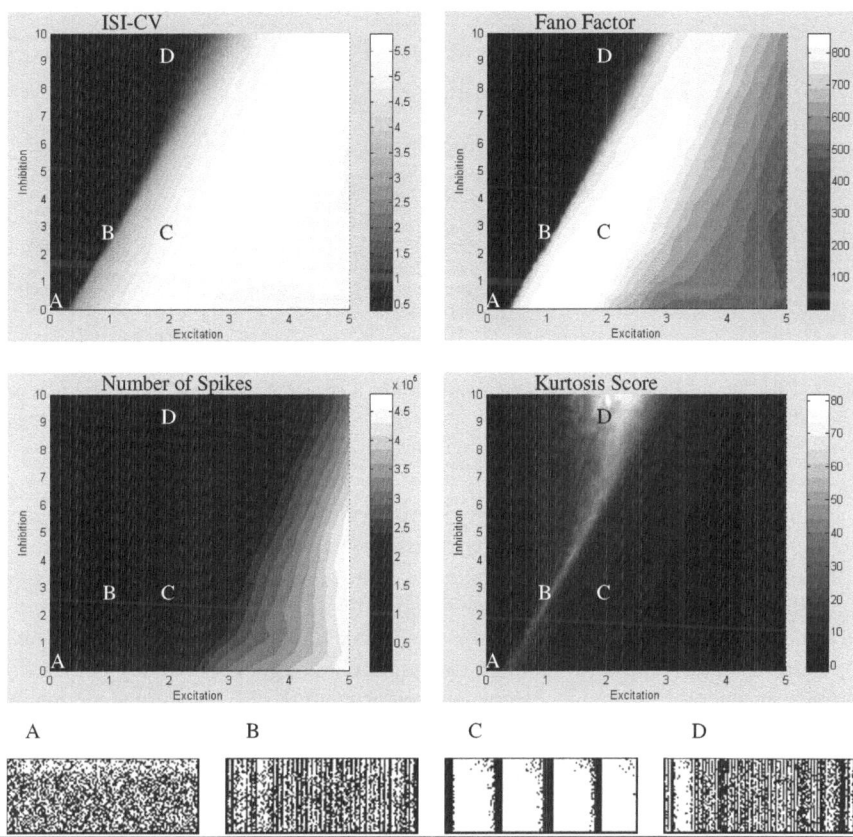

Fig. 2. Different measures of spike train characteristics, plotted across the excitatory and inhibitory weight space. The legend at right of each graph shows the shade correspondence to the displayed metric. The four networks from Figure 1(A-D) are placed in context on this graph. *(Top Left)*: ISI-CV is the standard deviation of the interspike interval divided by the mean. This value is maximized when the network is in a fully synchronized (seizure) state. This state is comprised predominantly of short periods of no activity punctuated by very long bursts of activity when all neurons fire almost simultaneously. *(Top Right)*: Fano Factor is similar in shape and analysis to ISI-CV except that its value is maximized when the network is in a seizure state comprised of *long* periods of no activity punctuated by very *short* bursts when all neurons fire almost simultaneously. *(Bottom Left)*: The total number of spikes generated by the network correlates with both Fano Factor and ISI-CV. *(Bottom Right)*: The Kurtosis Score identifies those regions in parameter space where slow oscillations predominate but are punctuated by short periods of fast oscillations (in the sampled networks, these fast oscillations occur at the slow oscillation peaks but this doesn't affect the score calculation – see Discussion). The slow oscillations give fewer spikes per millisecond and cause the peak of the distribution close to zero. The fast oscillations that cause many simultaneous or near-simultaneous spikes create the long tail. The combination of tall peak and long tail increases the kurtosis and, because the distribution is heavily skewed to the right (see histograms in Figure 1B and D) also increases the skewness. Seizure-like spike trains do not have this characteristic since they are comprised of two quite distinct peaks in the distribution, dramatically lowering the kurtosis.

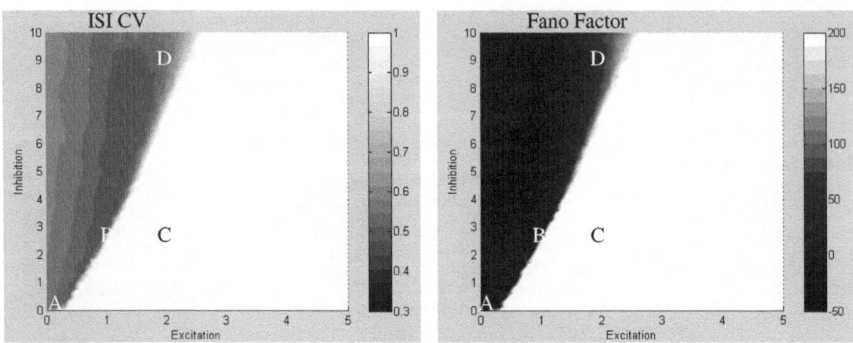

Fig. 3. ISI-CV and Fano Factor graphs from Figure 2 rescaled to reveal characteristics in the region of high kurtosis. ISI-CV decreases, while Fano Factor increases monotonically towards the seizure state throughout the region of high kurtosis.

High Kurtosis Score occurs in a region of weight space immediately adjacent to seizure-like activity (Figures 2 and 3). Thus to operate in a high Kurtosis Score regime the nervous system needs to maintain tight control over excitation and inhibition. Slightly too much of the former or slightly too little of the latter will tip the network into seizure. Given the lifetime prevalence of epilepsy of 2-5% in the general population [8], the failure of the brain to maintain this tight control may not be an uncommon problem. Continuous electro-encephalogram (EEG) recordings are available from epileptic patients before, during and after seizures, and it would be interesting to calculate these metrics on EEG data recorded prior to and during seizures to test for any diagnostic or predictive capabilities.

The power spectra of EEG traces over long time frames demonstrate a power law distribution [9]. The mammalian cortex also appears to be wired using small world connectivity principles, so that it simultaneously minimises path lengths and the required number of connections (see [10] for a synthesis of ideas on this topic). It seems the brain is using scale-free principles for both physical structure and macro-scale dynamics. Is it possible that micro-scale spiking behaviour may also be governed by similar principles? Intriguingly, it has been said that kurtosis should be thought of as the scale-free movement of probability from the shoulders of a distribution to the centre and tails [11]. Spike count distributions plotted on log-log axes show a tendency to straight lines for those spike trains with high Kurtosis Score (see Figure 1 insets: compare the curve of Figure 1A inset showing low Kurtosis Score with the more linear relationship in Figure 1D inset with the highest Kurtosis Score). While a linear relationship on a log-log graph is not sufficient to demonstrate a scale-free distribution, it is consistent with one, and it remains as future work to investigate the potential link. Note that for any scale-free distribution, the kurtosis will increase as the exponent in the power law (i.e. the slope of the line in the log-log plot) increases, whereas a true power law regression should be independent of slope.

Because Kurtosis Score is based on the spike count distribution, it omits all information about the *order* of the events in the spike train. An alternative spike train to

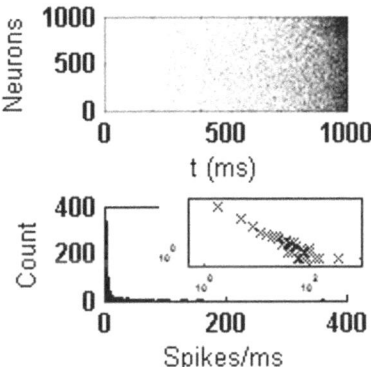

Fig. 4. A transformed spike train with the same Kurtosis Score and Fano Factor as Figure 1D. This spike train was obtained by dividing the raster plot of Figure 1D into 1 ms time slices, and sorting the slices according to spike count. The Kurtosis Scores for the original and transformed raster plots are the same since the spike count distribution is unchanged by the sorting process. The difference in the spike trains also shows that the Kurtosis Score gives very different information to a spectrum analysis.

that shown in Figure 1D, where the events are sorted in increasing number of spikes per millisecond, shows very different structure (see Figure 4). This spike train has the same Kurtosis Score as in Figure 1D because it has the same spike count distribution. Therefore to say that the Kurtosis Score identifies nested oscillations is incorrect; but in the context of spike events in networks of neurons, highly kurtotic distributions are much more likely to be generated by nested oscillations such as illustrated in Figure 1D than by spike trains like that shown in Figure 4. Also, true scale-free behaviour should be scale-free over a broad range of timescales (Figure 1D) rather than being dependent on the chosen sampling start and end times (Figure 4).

Just as the Fano Factor was developed in the context of the statistical properties of ionising radiation [4], yet is applicable in diverse domains, so too is Kurtosis Score. Events in any domain that occur discretely in time or space are amenable to Kurtosis Score calculation, and it will be particularly relevant where a power law relationship is hypothesised or known to exist. This includes graph and queuing theory, medicine and epidemiology, physics, geology, economics, ecology and sociology to name a few [12]. We also suggest that the technique used in this study of performing parameter sweeps across multi-dimensional spaces and displaying the results as heat maps can be a great aid for visualisation and understanding of complex system behaviour, and with the computing power now available this technique can be successfully applied to the simulated neural network domain.

Acknowledgements. The authors thank Markus Diesmann and Tom Tetzlaff for insightful discussions. This work was supported by an ARC Thinking Systems grant and a COSNet Overseas Travel grant (COSNet Proposal number 92).

References

1. Buzsaki, G.: Large-scale recording of neuronal ensembles. Nature Neuroscience 7(5), 446–451 (2004)
2. Hoagland, H.: Rhythmic Behavior of the Nervous System. Science 109(2825), 157–164 (1949)
3. Brunel, N.: Dynamics of Sparsely Connected Networks of Excitatory and Inhibitory Spiking Neurons. Journal of Computational Neuroscience 8(3), 183–208 (2000)
4. Fano, U.: Ionization Yield of Radiations. II. The Fluctuations of the Number of Ions. Physical Review 72(1), 26–29 (1947)
5. Werner, G., Mountcastle, V.B.: The Variability of Central Neural Activity in a Sensory System, and its Implications for the Central Reflection of Sensory Events. Journal of Neurophysiology 26(6), 958–977 (1963)
6. Izhikevich, E.M.: Simple model of spiking neurons. IEEE Transactions on Neural Networks 14(6), 1569–1572 (2003)
7. Abbott, L.F., et al.: Synaptic Depression and Cortical Gain Control. Science 275(5297), 221 (1997)
8. Sander, J.W.: The epidemiology of epilepsy revisited. Curr. Opin. Neurol 16(2), 165–170 (2003)
9. Linkenkaer-Hansen, K., et al.: Long-Range Temporal Correlations and Scaling Behavior in Human Brain Oscillations. Journal of Neuroscience 21(4), 1370 (2001)
10. Buzsáki, G.: Rhythms of the Brain. Oxford University Press, USA (2006)
11. Balanda, K.P., MacGillivray, H.L.: Kurtosis: A Critical Review. The American Statistician 42(2), 111–119 (1988)
12. Bak, P.: How nature works, Copernicus New York, NY, USA (1996)

Pioneeristic Works on Neuronal Nets:
A Short History

Maria Marinaro

Deptartment of Physics, University of Salerno Italy,
INFM, INFN Gruppo Coll di Salerno
International Institute for Advanced Scientific Studies "Eduardo R. Caianiello" (IIASS)
Vietri sul Mare (SA) – Italy
iiass.direzione@tin.it

Abstract. A short history of simulation models for living organism behaviour is presented. The machine designed by Bent Russell, 1913, to simulate the theory on nervous conduction, introduced by the psychologists Max Meyer and Edward Thorndike, is described. The new ideas on the mechanization of thought processes, carried out in the Cybernetic Age are commented and some models of neuronal nets based on these ideas are reported. Finally, a system of coupled oscillators, introduced by V. Braitenberg, E.R. Caianiello, F. Lauria and N. Onesto, is presented as an example of a cybernetic model comparing it with today's researches on nervous systems.

1 Pre-History – Russell's Machine

The branch of science that we now call Neuronal Network (n.n.) has been developed in the last sixty years. It is based on the works carried out by psychologists, engineers, physicists and mathematicians during the first half of the 20th century, works that opened the door to a new science, i.c. the Cybernetics, the science of communication and control.

At the beginning of the last century, the studies on organic and non-organic worlds were completely separated. The prejudice that capacities, as learning, recognition and so on, were distinctive features of living organisms ruled.

According to psychologists, man cannot build, using physical principles, machines showing capacities of living organisms.

But after a few years things started to change and some scientists began to consider the possibility to fill the gap between the organic and inorganic worlds. Some capacities of living organisms, as learning, were not viewed as the opposite of a mechanic behaviour but, on the contrary, as a particular kind of automatization.One step in this direction was made by the engineer S.B. Russell [18] (1913) who designed a machine to simulate certain hypothesis on the plasticity of nervous connections, pointed out by the psychologists M. Meyer and E.L. Thorndike.

Specifically, Meyer [12] formulated the *Drainage Theory of Nervous Conduction* "nervous flux runs like a fluid through a system of pipes of modifiable capacity connected by one-way valves (synapsis)", while Thorndike proposed the *Theory of the Reinforcements of Stimulus – Response (S-R) Connections.*

M. Marinaro, S. Scarpetta, and Y. Yamaguchi (Eds.): Dynamic Brain, LNCS 5286, pp. 123–130, 2008.
© Springer-Verlag Berlin Heidelberg 2008

The relationship between the plasticity of a nervous system and its learning capacity suggested in the works of Meyer and Thorndike was exposed explicitly by Russell through the hypothesis that continuous or frequent stimulations of neurons at short time intervals could result in a strengthening of the connections between neurons and an increase of their conductivity. While discontinuous or not frequent stimulations could weaken the connections and decrease conductivity.

Russell, on the basis of this nervous system image, designed and built a hydraulic machine that embodied the neurological hypothesis.

The fundamental unit of the machine is the transmitter, a valve with an inlet pipe to introduce a flux of air or water, and an outlet pipe to discharge it. The maximum opening of the transmitter changes in time and thus the quantity of the flux that is discharged.

The transmitter is a mechanical device which has a kind of memory, indeed it modifies its own behavior according to its previous functioning, its history. The transmitter simulates the nervous conductivity, conjectured by the psychologists, which is increased or decreased according to the frequency of the stimulus acting on the neurons.

The behavior of Russell's machine depends on what could be called the "experiences" of the transmitters.

The importance of the result obtained by Russell was recognized by the psychologist Meyer who pointed out that: if a machine is capable of learning and forgetting, using only the rules of mechanics, there is no necessity to invoke non-physical principles to account for the ability, typical of living organisms, to adapt to environments and to learn.

In conclusion, although only simple kinds of learning could be simulated at that time, the learning machines share with the living organisms some "essential elements" that reveal a common functional organization and justify the mechanicistic explanation of learning.

At the end of his work, Russell called for the importance of a cooperative work between scientists of different branches to develop researches on the nervous system. Unfortunately, Russell's appeal was not taken into consideration. At that time, multidisciplinary studies were not well considered and the mechanicistic interpretation of living organisms was accepted only by very few scientists.

2 The Cybernetic Age

Thirty years later Russell's appeal was appreciated. This happened in 1943 when two papers were published:

1) Behaviour, purpose and teleology;
2) A logical calculus of the ideas immanent in the nervous activity.

These papers opened a new age in the history of mechanization of thought processes, the so called Cybernetic Age, even if the word Cybernetic was introduced some years later (1947) by N. Wiener.

According to Wiener, Cybernetics, from the Greek word kybernetes (pilot – steersman), is a new science; the science that studies the laws of control and communication in animals and machines. You see, finally, the gap between organic and inorganic worlds seems overcome.

The new science accepts the mechanicistic interpretation of living organisms and goes beyond the deterministic evolution of the microscopic physics systems.

It takes into account the following points:

1) What is relevant in the systems is the common organization that they show and not the specific nature of their components. This statement was put forward in the paper "Behaviour, Purpose, Teleology" by A. Rosenblueth, N. Wiener, and J. Biglelov [16] to clarify that cybernetics is interested in the study of the function of systems independently from their composition.

2) The analytic method, that studies systems dividing them into elementary parts, cannot describe the cooperative behaviour of systems containing a large number of elements interacting with each other. The collective behaviour of a system shows characteristic "properties" that are not present in the sum of its single elements (Wholeness, Gestalt: The well known global perception that in psychology is called "structuralism").

3) The central characteristic of living organisms, whose evolution is fixed by the future, the finality, the aim of the action, cannot be described using the concept of energy. Energy and its transformations lead to the deterministic evolution in which everything is fixed by the past. In a science that deals with living organisms, deterministic evolution has to be substituted with probabilistic one; i.e. using the language of physics, we have to replace the Classical Mechanics with the Thermodynamic and substitute the role of Energy with that of Entropy. In this way the irreversibility of time, present in the organic world, is introduced naturally.

The relevance of Entropy in Cybernetics, the science of transmission and communications, finds its justification in the work of C.E. Shannon [19].

In 1948 Shannon, to put on rigorous bases the study of signals transmission in the telephone channels, defined a quantitative measure of the information. The information introduced by Shannon for a code of N elements takes the following expression:

$$ I = -\sum_{i=i}^{N} P_i \lg P_i \tag{1} $$

P_i is the probability of finding the element i of the code and N are the elements of the code. The quantity I is just what the physicists identify as the Entropy of the Code.

Based on these new ideas, at the beginning of the Cybernetic Age, many machines, robots, were designed and built, but the large part of these artifacts mimicked animal behaviour without illuminating the underlying functional principles that could justify the analogy between organisms and machines. The single devices gave the opportunity to explore the principles of automatism and control and were more engineering than psychological experiments.

But some years later, (1950), designing suitable machines, it was possible to test some theoretical hypotheses formulated by psychologists. In these cases the robots were material models embodying assumptions expressed in theoretical models.

Starting from the second part of the 1950s this requirement was at the core of several cybernetic programs. These programs were developed using different approaches according to the professional background of the scientists who developed them.

3 Theoretical Models: Neuronal Networks

The theoretical models introduced at that time had a common structure which imitate that of the nervous system: a set of elementary unities of elaboration interacting with each other through connections. To stress the analogy with the nervous system the unity and the connection were called respectively artificial neuron and synapsis.

The aim was to see whether, starting with very simplified models, it was possible to reproduce some intelligent actions as classification, recognition and learning. These models can be considered as the first examples of Neuronal Nets.

In 1943 W.S. McCulloch [11] and W. Pitts realized the first neuronal nets. Their models were based on the logic and the psychology available at the time. The models described the brain functions, by using logical calculus. Assembling a few neurons, suitably connected, McChulloch and Pitts designed devices that implemented logical functions. The impact of these results was enormous: any definition or statement corresponds to a network of neurons culminating in a single neuron that uniquely represented the logic preposition. Very important is that by means of this scheme it was possible to clarify the law of human behaviour.

McCulloch and Pitts' neuronal nets were able to reproduce logical calculus but could not learn. The plasticity present in any nervous system, adaptation to the environment, was completely absent in their networks. The learning ability was introduced some years later.

Two types of learning were developed:

1) Supervised learning;
2) Unsupervised learning.

Unsupervised learning was based on the ideas of the psychologist Donald Hebb [17] who, inspired by his studies on animal behaviour, proposed in 1942, the first learning rule for neuronal nets, the so called associative learning: "The coupling (synapsis) between two neurons is variable, it increases any time that the two neurons are simultaneously active".

Supervised Learning performed the update of the synapsis by requiring the network to give exact answers to a set of examples (training set).

The supervised learning was introduced later, by F. Rosenblatt [15], who proposed a general class of networks called Perceptrons in 1962.

The Perceptron is a one layer forward neuronal net that adapts itself to the environment using a learning algorithm. The algorithm changes the value of the synapsis between the input and the output neurons, any time the answer of the neuronal nets to the input example is wrong.

It was possible to show that after a period of training, the network was able to generalize giving correct answers to examples which were not contained in the training set.

In the same years B. Widrow and M.E. Hoft [21] proposed a different learning algorithm, the so called *ADALINE* (Adaptive Linear Neuron) or δ rule to update the synapsis of the perceptron. The new learning algorithm, based on the minimization of a suitable cost function, was very effective.

In 1961 E.R. Caianiello [4] published, in the first issue of the *Journal of Theoretical Biology* the work "An Outline of a Theory of Thought Processes and Thinking Machines".

The major innovations introduced in the paper were related to Caianiello's scientific background, who was a theoretical physicist with a vast experience in the study of complex systems.

He observed that in a nervous system there were two different scales of time: a fast one (10^{-3} sec.) that described the dynamics of the neuron and a very slow one (10^3 sec.) that controlled the adaptation of the nervous system to the environment, i.e. the change of the synapsis.

On the basis of this observation he divided the study of neuronal nets into two parts: dynamic and learning. Specifically, Caianiello assumed that in presence of two so different time scales, it is justified to study the evolution of neurons keeping the synapsis fixed and thus, extending to the neuronal nets the adiabatic hypothesis, well known in physics, he wrote two types of equations: The Dynamic and the Mnemonic Equations.

The Dynamic Equations describe the evolution of an ensemble of N neurons interacting each other with fixed coupling (synapsis). In formulae

$$\vec{X}_{t+\tau} = segn\left[\hat{A}\vec{X}_t + \vec{b}\right] \tag{2}$$

$\hat{A}$ is the NxN synapsis matrix independent on time, $\vec{X}_t$ is a vector with N components which describes the set of neurons at the time t. $\vec{b}$ is the threshold vector.

The Mnemonic Equations describe the evolution of the synapsis i.e. the learning processes related to the adaptation of neuronal nets to the environment. The mnemonic equations are a generalization of the Hebb rule:

The coupling between i and j neurons increases if the neuron i is active at the time t and the neuron j to time $t+1$, otherwise decreases. A threshold T limits the maximum value taken by the coupling. A schematic behaviour of the coupling $A_{k,h}$ is reported in the picture:

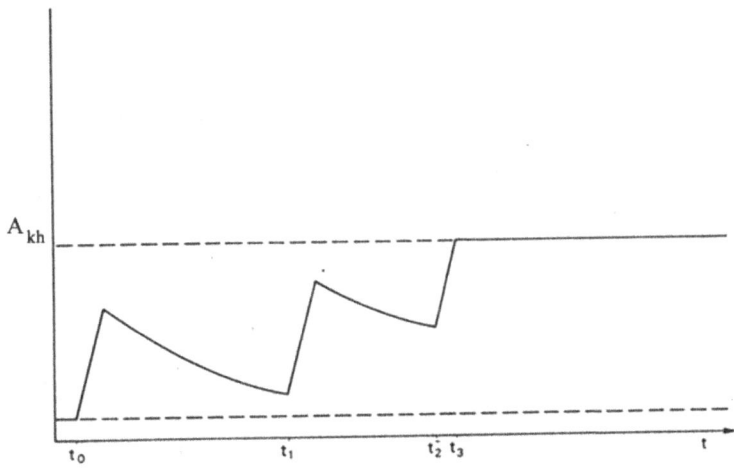

It is not worth explaining Caianiello's model in detail, but one should like to stress two points:

1) The model gives a remarkable contribution by introducing a solid frame of modellization into the study of nervous systems according to standard theoretical physics;
2) The adiabatic hypothesis was very important for the future development of the neuronal net models.

4   XOR Problem and Back Propagation

The researches on neuronal nets, developed so rapidly after the introduction of McCulloch's artificial neuron, were strongly depressed when in 1969 the book "Perceptron" by M.L. Minsky and S.R. Papert [13].

Minsky proved that the implementation of one of the elementary logical functions, the *XOR,* and then the possibility to use forward neuronal nets to perform logical calculus, cannot be performed by one layer network, but requires the use of multilayer structures.

Minsky and Papert's criticism and the lack, at that time, of a learning algorithm capable of updating multilayer neuronal nets produced much skepticism on the neuronal net computation capacities and a brusque decay of the scientific interest in this research field.

Looking back at that crisis some years later, it was evident that the loss of interest in the neural net researchers was due more to the outstanding position of Minsky and Papert, members of the Parallel Distributed Processes (PDP) of San Diego University in California, than to serious scientific reasons. Indeed, after the publication of the M.P. book the U.S. National funds for researchers in neural nets abruptly decreased, and for almost 20 years most of the computer science community left the neural net program.

Nevertheless, a few scientists continued to work on the topic and some important unsupervised models, based on the generalization of Hebb mechanisms or competitive mechanisms, were proposed in the following years.

Examples of these activities are:

1) The biologically plausible models introduced by S. Grossberg [6];
2) The Content Addressable Memory (CAM) by J.A. Anderson [2];
3) The Adaptive Resonance Theory (ART) by G.A. Carpenter and S. Grossberg [5];
4) The self Organization Map (SOM) proposed by T. Kohonen [9];
5) The Topographic Map by S.A. Amari [1];
6) The model introduced by J.J. Hopfield [8] to implement associative memory which is an interesting reformulation of the Content Addressable Memory (CAM) in terms of physical systems. The model became very popular and helped to recall the attention on neural net research.

More details on this period can be found in the book: *Neurocomputing: Foundations of Research,* Cambridge 1988, MIT Press.

In 1986, a new age on the research of neuronal nets began. This happened when D.E. Rumelhart and J.L.McClelland [17] proposed the Back Propagation Algorithm to update the synapsis of a multilayer neuronal net. The new algorithm, an extension of the δ Rule by Windrow, already introduced by Y. Le Cun [10], D.B. Parker [14], and P. Werbos [20] in different contexts, overcame Minsky and Papert's criticism regarding the use of neuronal nets to perform logical calculus and raised a new interest in this research field.

Applications of neuronal net algorithms in different branches: classification, pattern-recognition, control, learning, were performed and comparison with other methodologies, as the standard statistic approach, fuzzy sets and so on, put in evidence the points of stress and weakness of the neuronal net algorithm.

Today, models containing new neurobiological plausible mechanisms have been introduced and the use of neuronal nets to clarify neuroscience problems is of big interest.

5 An Array of Coupled Oscillators to Model Neuronal Behaviour

In conclusion we would like to mention the paper "A System of Coupled Oscillators as a Functional Model of Neuronal Assemblies" published in 1959 in the journal *Il Nuovo Cimento*. The paper was written by V. Braitenberg (neuroscientist), E.R. Caianiello (physicist) and two mathematicians F. Lauria and N. Onesto [3].

We have selected this paper for two reasons:

1) To present a witness of the interdisciplinary activity on neuronal nets carried out in Naples by the E.R. Caianiello group since 1956;

2) To illustrate a modellization of nervous functions which satisfy some principles that Wiener, ten years earlier, had set at the basis of the new science "Cybernetics".

The proposed network is very modern and up-to-date.

Specifically, in the paper, a simple model, based on a neurobiological plausible mechanism, is presented: an array of coupled oscillators. The coupling between oscillators is variable. Inputs, represented by the changes of some couplings of the array, generate outputs that show different frequency spectra of the total oscillations of the network.

a)   The transmission of the signals, the input-output relations, are produced by a cooperative behaviour. The system realizes the wholeness effect (Gestalt); Indeed, the output modes of the network are typical of the system as a whole and are different from the modes of single oscillators. The oscillation of the array depends on the couplings and not on the state of the single unities.

b)   The network shows learning capacities. An example of associative memory is realized: the input-output acquires the basic properties of the recall;

c)   A non localized memory is realized in the network. The record of the input patterns is delocalized, it is distributed on couplings which are in different sites of the array.

References

1. Amari, S.A.: Topographic Organization of Nerve Fields. Bulletin of Mathematical Biology 42, 339–364 (1980)
2. Anderson, J.A.: A Memory Model Using Spatial Correlation Functions. Kybernetik 5, 113–119 (1968)

3. Braitenberg, V., Caianiello, E.R., Lauria, F., Onesto, N.: A System of couplet Oscillators as a Functional Model of Neuronal Assemblies. Il Nuovo Cimento 11, 278–282 (1959)
4. Caianiello, E.R.: Outline of a Theory of Thought-Processes and Thinking Machines. Journal of Theoretical Biology 1, 204–235 (1961)
5. Carpenter, G.A., Grossberg, S.: A Massively Parallel Architecture for Self-Organizing Neural Pattern Recognition Machine. Computer Vision, Graphics, and image Processing 37, 54–115 (1987)
6. Grossberg, S.: Competitive Learning: from interactive Activation to Adaptive Resonance. Copitive Scide II, 23–63 (1987)
7. Hebb, D.O.: The Organization of Behavior. Wiley, New York (1949)
8. Hopfield, J.J.: Neural Networks and Physical Systems with Emergent Collective Computational Abilities. Proceedings of the National Academy of Sciences 79, 2554–2558 (1982)
9. Kohonen, T.: Self-Organized Formation of Topologically Correct Feature Maps. Biological Cybernetics 43, 59–69 (1982)
10. Le Cun, Y.: Une Procédure d'Apprentissage pour Réseau à Seuil Assymétrique. In: Cognitiva 1985: A la Frontière de l'Intelligence Artificielle des Sciences de la Connaissance des Neurosciences, pp. 599–604. CESTA, Paris (1985)
11. McCulloch, W.S., Pitts, W.: A Logical Calculus of the Ideas Immanent in Nervous Activity. Bull. Math. Biophys. 5, 115–133 (1943)
12. Meyer, N.: The Comparative Value of Various Conceptions of Nervous Function Based on Mechanical Analogies. Am. J. Psychol. 24, 555–563 (1913)
13. Minsky, M.L., Papert, S.A.: Perceptrons. MIT Press, Cambridge (1969)
14. Parker, D.B.: Learning Logic, Technical Report TR-47, Center for Computational Research in Economics and Management Science, Massachusetts Institute of Technology, Cambridge, MA (1985)
15. Rosenblatt, F.: Principles of Neurodynamics, New York Spartan (1962)
16. Rosenblueth, A., Wiener, N., Bigelow, J.: Behavior, Purpose and Teleology. Philos. Sci. 10, 18–24 (1943)
17. Rumelhart, D.E., McClelland, J.L., PDP Research Group: Parallel Distributed Processing: Explorations in the Microstructure of Cognition. Foundations, vol. 1. MIT Press, Cambridge (1986)
18. Russel, S.B.: A Practical Device to Simulate the Working of Nervous Discharges. J. Animal Behav. 3, 15–35 (1913)
19. Shannon, E.C.: A Mathematical Theory of communication. Bell System Technical Journals 27 (1948)
20. Werbos, P.: Beyond Regression: New Tools for Prediction and Analysis in the Behavioral Sciences, Ph.D Thesis, Harvard University (1974)
21. Widrow, B.: Generalization and Information Storage in Networks of Adaline "Neurons". In: Yovits, M.C., Jacobi, G.T., Goldstein, G.D. (eds.) Self-Organizing Systems, Washington Spartan (1962)
22. Wiener, N.: Cybernetic, or control and Communication in the Animal and the Machine. Wiley, New York (1948)

Place-Field and Memory Formation in the Hippocampus

Gergely Papp

SISSA – Cognitive Neuroscience, via Beirut 2, 34014 Trieste, Italy
papp@sissa.it

Abstract. CA3 region of hippocampus may be considered as an initial store-room for memories. During spatial memory tasks CA3 cells were found to fire in single well defined portions of a recording room (place-fields), whereas cells in both of its afferent regions, entorhinal cortex (EC) and dentate gyrus (DG), showed activity at multiple locations within an environment. DG by prior theoretical studies was proposed to be a "teacher", by dominating activity through its strong and very sparse connectivity during storage in CA3. It becomes however an intriguing question, how DG, firing at multiple locations, may set up a new spatial memory composed of exclusively single place-fields in CA3. Here we report that dentate gyrus is necessary to set up novel spatial memories in CA3 and multiple peaks of dentate gyrus may combined into mostly single fields by competitive learning in CA3.

Keywords: Hippocampus, dentate gyrus, mossy fibers, competitive learning.

1 Introduction

Marr formulated a theory how hippocampus may form new memories [1]. The most important assumption of his work was that recurrent connections in CA3 may play a crucial role in memory recall, during which a noisy version of a stored memory via reverberation through the CA3 collaterals would be restored to the full representation. A mathematical model by Hopfield [2] based on Marr's idea identified memory to be the attractor states of such networks, with the maximum number of retrievable stable states scaling with the number of associatively modifiable recurrent connections per cell [3].

There is, however, a conflict between storage of new patterns and recall of previous memories. During storage the network has to be driven by external stimuli, as opposed to recall when activity should rather be determined by the recurrent collaterals. Further, storage of correlated patterns would decrease number of memories that can be recalled. As proposed by McNaughton and Morris [4], mossy fibers, in their model taken to be a strong one-to-one connection between DG and CA3, may effectively suppress recurrents during storage and transfer the activity of DG to CA3 region, to the associative memory. As clarified by Treves and Rolls [5], what matters, instead, is that mossy fiber synapses be strong, sparse and relay sparse activity from DG, where codes for different memories could be decorrelated by various mechanisms. This would be sufficient to select a limited ensemble of CA3 cells coding for a new memory. The theory thus far is backed by experimental results [6]. Storage

M. Marinaro, S. Scarpetta, and Y. Yamaguchi (Eds.): Dynamic Brain, LNCS 5286, pp. 131–136, 2008.

mediated by DG may be coupled with neuromodulation found to selectively suppress recurrent collaterals and enhance synaptic plasticity during storage [7,8].

In spatial memory tasks, as a rat forages in an environment, hippocampal CA3 and CA1 cells tend to fire in a single limited portion of physical space, emitting virtually no spikes elsewhere [9]. Input to hippocampus arrives via the entorhinal cortex [10], where, in its dorsal part, cells were found to fire at multiple locations within one environment, arranged on the vertices of a triangular grid [11]. Interestingly, nearby cells were found to fire with the same spatial frequency and orientation, whereas neurons, located further apart showed a different orientation and spatial frequency increased towards more ventral locations. Cells in DG were also found to fire in multiple peaks [12], lacking however a clear spatial organization as described for entorhinal cortex.

Firing correlates of both mEC and DG cells, contacting CA3 cells directly, was thus found to be strikingly different from place-fields, it is thus an intriguing question, how place-field firing of CA3 cells could be understood by the newly discovered firing correlates of its afferents.

Early models concentrated on explaining formation of place-fields via perforant path only [13-15]. A model by Solstad et al [13] shows single peaks would be obtained only if CA3 cells would integrate exclusively from those EC cells that have a peak in common in the same position where the CA3 cell has its place-field. Such a selection may be mediated by diverse mechanisms. Indeed, competitive learning [14] may decrease the number of resulting place-fields, as well as phase-precession [15], accounting, however, mainly for place-field formation in area DG.

In order thus to understand formation of place-fields in CA3, recurrent connections have also to be considered. An effective storage, as reviewed already [4,5], would necessitate strong inputs from DG. Also, spatial memories seem to represent a marginally stable state and are susceptible to noise [19], hence a clear spatial metric, decreasing "wrinkles" on these maps, during training is needed, transmitted through mossy fiber synapses. Recurrent collaterals then, if appropriately trained, could decrease more efficiently, than a network without a dentate area, the number of firing fields in region of CA3 as well as recall stored maps better. Given, however, that mossy fiber synapses, thought to set up new memories in CA3, were shown to have multiple peaks, as well as there is a slight convergence from few but multiple dentate cells towards CA3, it is unclear, how such inputs would finally combine to single peaks in CA3.

2   Methods

To find an answer to these question, we simulated a simplified hippocampal network comprising of medial entorhinal cortex and CA3, with and without a region simulating dentate granule cells, active only during training. A simplified dentate layer was used during training, where each cell fired at a single spatial location with one-to-one connections between DG to CA3. We first show that number of fields in CA3, compared to a network without a layer of DG, tend to be smaller, as a result of several iterations through appropriately trained recurrent connections. Next, we assumed multiple fields in DG and a slight convergence from several dentate cells in CA3. A place-field in CA3 may form where firing fields from DG coincide. Here we tested the hypothesis that competitive learning then may eliminate additional peaks, leading to the formation of single peaks in CA3.

In our model, as described in [20], we first assumed dentate cells to fire in single peaks as well as to have strong, one-to-one connections between DG and CA3, whereas contact probability between entorhinal cortex and CA3 cells was much higher. In the second part, dentate cells were taken to have double fields, and each CA3 cell received contacts from two dentate cells. Only medial EC cells, thus grid cells were simulated, with 10 orientation and spatial frequency bands ranging from 1 up to 3 peaks per environment. During training a model rat foraged in a simulated, 1x1 m square box, at each step weights between EC and CA3 as well as the recurrent collaterals in CA3 were updated using a modified Hebbian learning rule. During testing the model rat foraged in the same simulated environment, where at each position a partial cue was provided to EC and a template matching decoding was applied to test accuracy of network response. Templates were recorded from 400 positions within the environment, with input gradually fading to 0 in order to get representation stored in memory. Besides estimating memory performance of the network, number of firing fields were also counted.

3 Results

Simulating DG with single peaks and one-to-one connections was found to set up a stronger memory in CA3 than in a network without a dentate region, as shown by percent correct localization using template matching decoding (Fig 1A). Number of firing peaks in memory representation in CA3 decreased for both networks, but to considerably lower levels in case of using dentate cells, reaching bona fide single peaks after 3 epochs of training (Fig 1B; periodic boundary conditions were used). Without dentate gyrus several cells with double fields were found, shown also on 3 representative cells and several cells developed overly large fields (Fig1C). The reason for a worse performance on percent correct localization was due to a collapse to some final attractor states, due to a lack of strong metric information during training.

We next compared, in the case when dentate was simulated, performance of the network at pattern onset, corresponding to a purely feed-forward state, with that after each iteration in the recurrent connections. Number of place-fields (Fig 1E), evaluated after 2 epochs of training, decreased reliably during recall, that is, reverberation in recurrent connections further decreased the number of firing peaks in CA3, shown also on 3 example cells (Fig 1F). A decrease in number of fields was accompanied by a steady rise in percent correct performance, reaching a maximum after approximately the same number of iterations (8 iterations) when minimum number of place-fields were counted (7 iterations).

Next, to simulate a more realistic dentate gyrus, dentate cells were assumed to have double peaks with a slight convergence from two dentate cells on one CA3 cell during learning. During training, number of peaks decreased from an initially high number to

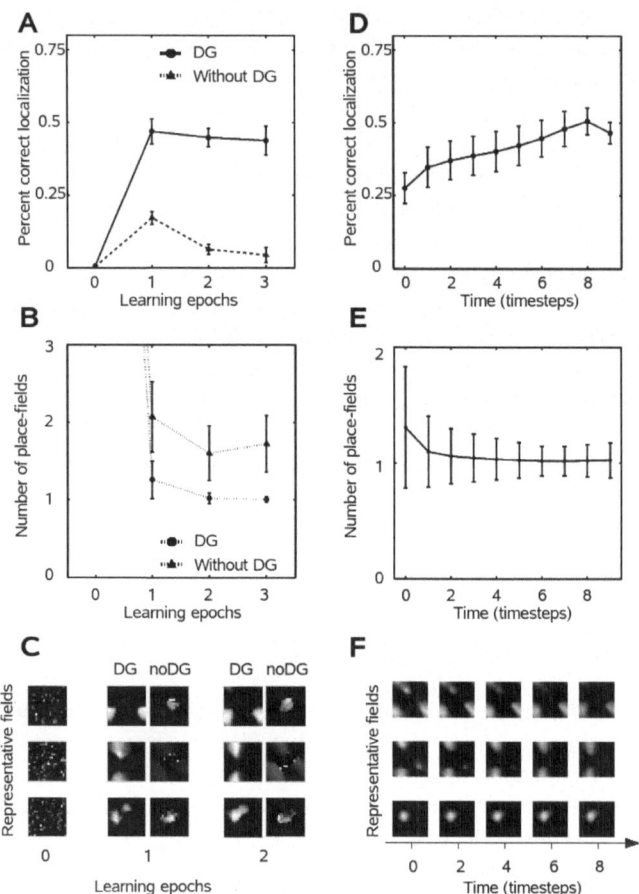

Fig. 1. Development of place-fields with and without a simulated DG area. (**A**) Percent correct localization increases during learning, but decays back in the absence of DG. (**B**) Enhanced decrement in number or place-fields was seen when DG was simulated. (**C**) Sample place-fields from networks with and without DG, after 9 iterations (roughly 100 ms) in the recurrent network. Compact place-fields are formed only when DG is simulated. Percent correct localization increases (**D**), whereas number of place-fields decreases (**E**) during reverberation in the recurrent network. (**F**) Development of place-fields during reverberation in the recurrent network, smaller secondary peaks tended to vanish. Error bars: s.d., except for B: s.e.m.

a level with many CA3 cells having single fields, thus significantly below a level expected if CA3 cells would fire in all positions where their afferent dentate cells were active (between 3 and 4). Firing patterns in CA3, however, did not reach the criterion of *bona fide* place-fields (Fig 2B). Percent correct performance (Fig 2A) was found to increase considerably due to training, the network, however, showed less accuracy recalling locations as did the network trained with a simplified dentate (cf. Fig 1A).

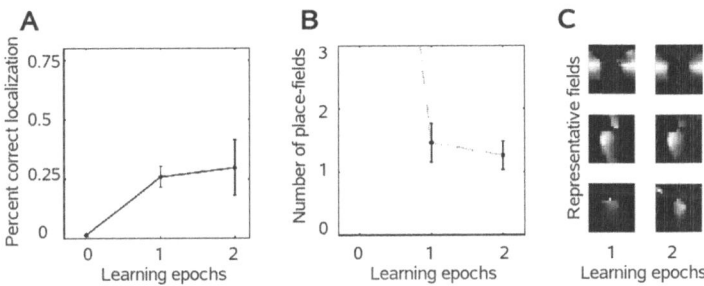

Fig. 2. Development of place-fields in CA3 trained with a more realistic DG. (**A**) Percent correct localization increases considerably, but to a lesser extent when considering single peaks in DG. (**B**) Number of firing peaks in CA3 decreases to near single fields in CA3, but a considerable fraction of cells fire in more than one location. (**C**) Representative cells from the first and second learning epoch, showing single and double fields. Error bars: s.d., except in B: s.e.m.

4 Discussion

Cells in medial entorhinal cortex projecting to hippocampus were found to fire in multiple locations within one environment, arranged on the vertices of a triangular grid [11] and cells in DG, thought to dominate activity in CA3 [4,5] during storage in order to set up new memories, were also found to fire in multiple, randomly positioned peaks [12]. Models, using feed-forward architecture could thus far account for formation of multiple peaks, generally found in DG [13-15].

Here we showed input from dentate gyrus is necessary to set up stable memories as well as to reach bona fide place-fields in an associative memory like CA3 region of hippocampus. Our model is also able to give partial account for the formation of single fields in CA3 when more realistic parameters for dentate cells are considered, thus multiple peaks together with a slight convergence from DG towards CA3. As a considerable portion of CA3 cells fired at double locations, other ingredients should also be considered, like synaptic modifications between DG and CA3.

References

1. Marr, D.: Simple memory: a theory for archicortex. Philos. Trans. R. Soc. Lond. B. Biol. Sci. 262, 23–81 (1971)
2. Hopfield, J.J.: Neural networks and physical systems with emergent collective computational abilities. Proc. Natl. Acad. Sci. 79, 2554–2558 (1982)
3. Amit, D.J., Gutfreund, H., Sompolinsky, H.: Statistical mechanics of neural networks near saturation. Ann. Phys (N.Y.) 173, 30–67 (1987)
4. McNaughton, B.L., Morris, R.G.M.: Hippocampal synaptic enhancement and information storage within a distributed memory system. Trends Neurosci. 10, 408–415 (1987)
5. Treves, A., Rolls, E.T.: Computational constraints suggest the need for two distinct input systems to the hippocampal CA3 network. Hippocampus 2, 189–199 (1992)

6. Lee, I., Kesner, R.P.: Encoding versus retrieval of spatial memory: double dissociation between the dentate gyrus and the perforant path inputs into CA3 in the dorsal hippocampus. Hippocampus 14, 66–76 (2004)

7. Hasselmo, M.E., Schnell, E.: Laminar selectivity of the cholinergic suppression of synaptic transmission in rat hippocampal region CA1: Computational modeling and brain slice physiology. J. Neurosci 14, 3898–3914 (1994)

8. Villarreal, D.M., Gross, A.L., Derrick, B.E.: Modulation of CA3 afferent inputs by novelty and theta rhythm. J. Neurosci. 27, 13457–13467 (2007)

9. O'Keefe, J., Dostrovsky, J.: The hippocampus as a spatial map. Preliminary evidence from unit activity in the freely-moving rat. Brain Res. 34, 171–175 (1971)

10. Dolorfo, C. L., Amaral, D.G.: Entorhinal cortex of the rat: topographic organization of the cells of origin of the perforant path projection to the dentate gyrus. J. Comp. Neurol. 398, 25–48 (1998)

11. Hafting, T., Fyhn, M., Molden, S., Moser, M.B., Moser, E.I.: Microstructure of a spatial map in the entorhinal cortex. Nature 436, 801–806 (2005)

12. Leutgeb, J.K., Leutgeb, S., Moser, M.B., Moser, E.I.: Pattern separation in the dentate gyrus and CA3 of the hippocampus. Science 315, 961–966 (2007)

13. Solstad, T., Moser, E.I., Einevoll, G.: From grid cells to place cells: a mathematical model. Hippocampus 16, 1026–1031 (2006)

14. Rolls, E.T., Stringer, S.M., Elliot, T.: Entorhinal cortex grid cells can map to hippocampal place-cells by competitive learning. Network 17, 447–465 (2006)

15. Molter, C., Yamaguchi, Y.: Entorhinal theta phase precession mechanism sculpts dentate gyrus place fields. Hippocampus (in press)

16. Rolls, E.T., Kesner, R.P.: A computational theory of hippocampal function, and empirical test of the theory. Progr. Neurobiol. 79, 1–48 (2006)

17. Papp, G., Treves, A.: Continuous attractors come fragmented. Soc. Neurosci. Abstract 68, 13 (2006)

18. Papp, G., Witter, M.P., Treves, A.: The CA3 network as a memory store for spatial representations. Learn. Mem. 14, 732–744 (2007)

Improving Recall in an Associative Neural Network of Spiking Neurons

Russell Hunter[1], Stuart Cobb[2], and Bruce P. Graham[1]

[1] Department of Computing Science and Mathematics, University of Stirling, Stirling, FK9 4LA, U.K.
{*rhu,b.graham}@cs.stir.ac.uk
[2] Division of Neuroscience and Biomedical Systems, University of Glasgow, Glasgow, G12 8QQ, U.K.
s.cobb@bio.gla.ac.uk

Abstract. The mammalian hippocampus has often been compared to neural networks of associative memory [6]. Previous investigation of associative memory in the brain using associative neural networks have lacked biological complexity. Using a network of biologically plausible spiking neurons we examine associative memory function against results for a simple artificial neural net [5]. We investigate implementations of methods for improving recall under biologically realistic conditions.

Keywords: Associative memory, mammalian hippocampus, neural networks, pattern recall, inhibition.

1 Introduction

Graham and Willshaw [5] examined the performance of pattern recall in an artificial neural network of associative memory. The net used simple binary units with 10% partial connectivity. They investigated methods to improve the quality of pattern recall using variations of the winners-take-all (WTA) approach. The WTA approach chooses the required number of units with the highest dendritic sum to fire during pattern recall. Using a model network of biologically realistic spiking neurons [8], we investigate the application of local inhibitory circuitry and a modification in the membrane properties of the PCs in an attempt to improve recall quality and replicate the methods of Graham and Willshaw [5].

2 The Model

The network model of autoassociative memory contains 100 recurrently connected cells (fig. 1a). The model is based on the network of Sommer and Wennekers [8]. It was created and simulated using the Neuron computer simulation package [1].

Each cell was an identical two-compartment model of a CA3 pyramidal cell, developed by Pinsky and Rinzel [7]. In each cell the compartments are coupled

M. Marinaro, S. Scarpetta, and Y. Yamaguchi (Eds.): Dynamic Brain, LNCS 5286, pp. 137–141, 2008.

electrotonically by g_c, the strength of coupling and p, the percentage of total area in the soma-like compartment. The soma-like compartment has fast sodium and potassium currents that can generate action potentials (AP). The dendrite contains slower calcium and calcium-modulated currents.

Each PC is connected to every other PC (non-reciprocal) with a probability of 0.1 (10% connectivity). Connections use an AMPA synapse which generates a fast excitatory post-synaptic potential. Synaptic delay varies from 0.3 to 1 ms, and peak conductance ranges up to $G_{AMPA} = 0.0154\mu S$. Higher conductance is required at lower levels of connectivity to maintain synaptic drive onto each cell. The actual connectivity is dependent upon the number of patterns stored, in combination with the physical connectivity.

A pattern consists of 10 randomly-chosen active neurons out of the population of 100. Patterns are stored by clipped Hebbian synaptic modification, resulting in a structured weight matrix. For a given pattern, the Hebbian rule specifies a weight of 1 for a connection between two neurons that are both active in the pattern, with all other weights being 0.

The network also contains global inhibition (fig. 1a) which acts as a threshold control that moderates the activity level in the network and keeps that activity from unphysiological states where all cells fire at very high rates [8]. It restricts firing rates to approximately gamma frequencies [2]. In the model, the inhibitory dynamics are not induced by explicit interneurons. It is assumed that APs of PCs envoke IPSPs on all cells in the network via inhibitory connections [8]. These inhibitory synapses employ a fast GABA-ergic conductance change with reversal potential $V_{CL} = -75\,mV$ and a fast rise-time and slow decay. The connection delay was around 2 ms. The inhibitory peak conductance was fixed at $G_{GABA} = 0.00017\mu S$.

Recall was tested by tonically stimulating 5 from a known pattern of 10 PCs using current injection to either the soma or dendrite with a strength ranging

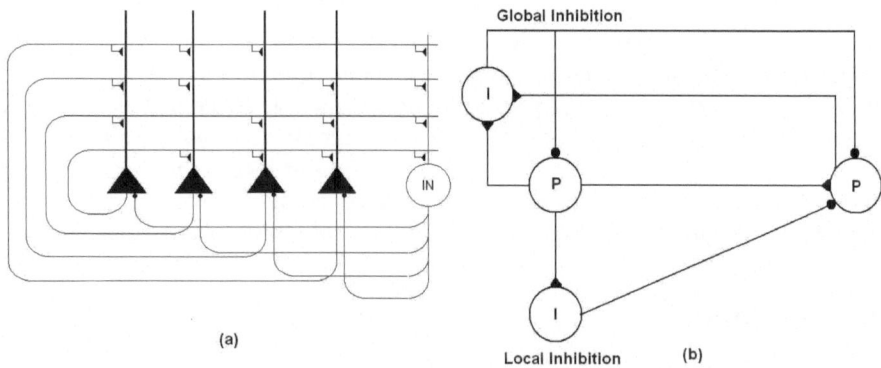

Fig. 1. (a) Circuit diagram of the network. PCs have an apical dendrite and soma. There are recurrent excitatory connections between cells but with no connections onto the same cell. Network is fully connected in this example. The IN cell represents the global inhibition mediated by activity from spiking PCs. (b) Normalised WTA with two PCs: basic global inhibition and a local inhibitory connection.

between 0.00075 and 0.0075 nA over 1500ms. Successful recall would result in the remaining 5 PCs in the pattern becoming active, but no other cells.

3   Thresholding Strategies for Recall

Standard winners-take-all (WTA). Pattern recall proceeds by only those PCs that receive the greatest excitatory input becoming active. In this network of spiking neurons, the standard WTA method recalls a pattern where the threshold of a PC's firing is set by the intrinsic threshold of the PC and the global inhibition. The intrinsic threshold of a PC is largely determined by membrane resistance and sodium channel density.

Normalised WTA network (localized inhibition). The normalised WTA uses the fact that all dendritic sums lie between a range 0 and some maximal level of input activity, which equates with the number of physical connections onto a cell that are active, irrespective of the learnt synaptic weight. This input activity is the amount of excitation each cell could receive, whereas the dendritic sum is the amount of excitation the cell actually receives. Graham and Willshaw [5] found that by normalising a cell's dendritic sum by its input activity reduces the error/overlap during recall. Local inhibition, implemented by having inhibitory connections between PCs corresponding to all possible modified excitatory connections in a partially connected net (fig. 1b), should have a similar outcome. This inhibition inhibits a PC in proportion to the excitation it could receive and could be considered as part of a disynaptic inhibitory drive with a fast acting GABAA type synapse [3].

Amplified WTA method. The average excitation a cell receives during recall increases with the cell's unit usage, leading to increasing overlap between the dendritic sums of high and low cells. Graham and Willshaw [5] found that for cells with a given unit usage, the variations/overlap due to unit usage can be reduced by a suitable transformation of the dendritic sum as a function of a cell's unit usage. Graham [4] used a method of signal (EPSP) amplification to help discriminate between low and high cells and therefore improve pattern recognition. Adding a persistent sodium channel to the soma with a low voltage activation range and appropriate maximum conductance should amplify high dendritic sums (summed EPSPs). Testing on a single cell shows a non-linear increase in dendritic summation above a given threshold.

4   Results

Recall performance was tested by storing 50 patterns in the net. The partial cue is 5/10 cells of a stored pattern. Physical connectivity was set at 10%. The dynamics of the network determines that the recall process is synchronous in which cell activation outwith the input cue is dependent on APs from the cued

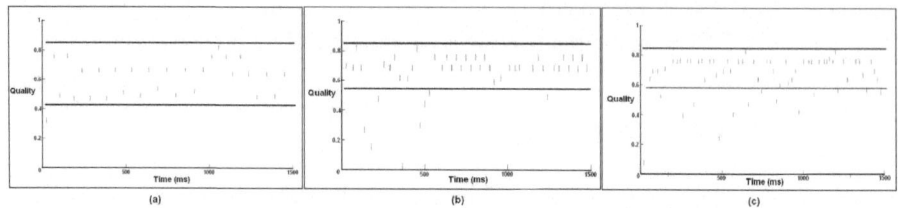

Fig. 2. Recall quality over time in a 10% partially connected network when recalling a single pattern using (a) standard WTA, (b) normalised WTA and (c) amplified WTA. Throughout, $I_d = 0.0075$ nA and $G_{GABA(g)} = 0.00017$ μS. (a) $G_{AMPA} = 0.0154$ μS; (b) $G_{AMPA} = 0.0154$ μS, $G_{GABA(l)} = 0.00748$ μS; (c) $G_{AMPA} = 0.008$ μS, $G_{pNa} = 0.000165$ μS. The horizontal lines are qualitative indicators of the main spread of recall performance in each case.

cells. The global inhibitory circuit synchronises the network activity to gamma frequency range.

Quality of recall was measured by examining spiking activity over 1500ms and calculating a quality measure as:

$$C = \frac{\sum_{i=1}^{N}(B_i - \bar{B})(B^* - \alpha_B)}{\left(\sum_{i=1}^{N}(B_i - \bar{B})^2 \sum_{i=1}^{N}(B^* - \alpha_B)^2\right)^{1/2}}$$

where C is the recall quality, B is a recalled output vector, B^* is the required output, N is the number of cells, α_B is the mean output activity and $\bar{B}$ is the mean activity of the recalled pattern. The required output vector is the selected memory pattern stored in the structured connectivity matrix. The actual output vector is determined by the APs from any cell occurring within a given sliding time window of 16 ms. This time was selected on the basis of spiking frequency during recall, so that at most a single spike from a cell would occur in the window.

With standard WTA the mean pattern recall quality is approximately 61% (fig. 2a). Using the normalised WTA (fig. 2b) the addition of localised inhibition improves the mean pattern recall quality to approximately 64%. A significant improvement can also be measured using the amplified WTA which gives a mean pattern recall quality over of approximately 65%. The low percentage of recall quality for each method suggests confusion from inherent noise due to overlap in patterns during the storage procedure and partial physical connectivity. The standard WTA approach (fig. 2a) shows an oscillation between high and low values of recall and a wide variation in the quality of pattern recall over time. The normalised WTA (fig. 2b) has a faster rate of cell spiking due to the localised inhibitory circuit. Also, the variation in recall quality is greatly reduced, with a range of 60% to 80% (excluding some outliers), compared to the standard WTA at approximately 40% to 80%. Similarly, the amplified WTA approach (fig. 2c) shows less variation in quality of recall per iteration with a range of 60% to 80% and fewer outliers. Outliers can be attributed to increased iterations from the extra inhibition in the normalised WTA method and the increased likelines of an

AP due to the persistent Na channel in the amplified WTA. The mean quality over all patterns shows a statistically significant (95% CI) increase when using the normalised and amplified methods compared to the standard WTA method.

5 Conclusion

Our model demonstrates that methods of improving recall in a network of spiking neurons show significant correlations to the results found in artificial neural networks of associative memory [5]. We have shown, as found experimentally [2], that global inhibition is required for synchronous PC activity in the gamma frequency range. Our model also suggests that for pattern recall, a method of local inhibition (GABA-ergic interneurons) may further synchronize the activity between PCs and also improve the recall of a pattern. Adding a persistent Na channel to the cell to amplify large EPSPs also improved the quality of pattern recall. This result suggests that the membrane properties of PCs may be able to reduce noise in patterns of synaptic input. The added persistent Na channel confirms the methods explored in [4], where it was found that voltage-gated ion channels act to boost synaptic input and improve recall in a model of associative memory.

Acknowledgement. This work was funded by an EPSRC project grant to B. Graham and S. Cobb.

References

[1] Carnevale, N.T., Hines, M.L.: The Neuron Book. Cambridge University Press, Cambridge (2005)

[2] Cobb, S.R., Buhl, E.H., Halasy, K., Paulsen, O., Somogyi, P.: Synchronization of neuronal activity in hippocampus by individual GABAergic interneurons. Nature 378, 75–78 (1995)

[3] Fransen, E., Lasner, A.: A model of cortical associative memory based on a horizontal network of connected columns. Network Comput. Neural Syst. 9, 235–264 (1998)

[4] Graham, B.P.: Pattern recognition in a compartmental model of a CA1 pyramidal neuron. Network Comput. Neural Syst. 12, 473–492 (2001)

[5] Graham, B., Willshaw, D.: Improving recall from an associative memory. Bio. Cybernetics 72, 337–346 (1995)

[6] Graham, B., Willshaw, D.: Capacity and information efficiency of the associative net. Network 8, 35–54 (1997)

[7] Pinsky, P., Rinzel, J.: Intrinsic and Network Rhythmogenesis in a Reduced Traub Model for CA3 Neurons. J. Comput. Neuroscience 1, 39–60 (1994)

[8] Sommer, F.T., Wennekers, T.: Associative memory in networks of spiking neurons. Neural Networks 14, 825–834 (2001)

Author Index